Rapid Assessment Program

Working

Papers

1

A Biological Assessment of the Alto Madidi Region

and adjacent areas of

Northwest Bolivia

May 18 - June 15, 1990

CONSERVATION INTERNATIONAL

DECEMBER 1991

Rapid Assessment Working Papers are occasional reports published three to five times a year. For subscription information write to:

Conservation International

Publications

1015 18th Street, NW

Suite 1000

Washington, DC 20036

Tel: 202/429-5660

Fax: 202/887-5188

Conservation International is a private, nonprofit organization exempt from federal income tax under section 501 (c)(3) of the Internal Revenue Code.

Library of Congress Catalog Card Number 91-078133

Printed on recycled paper

Table of Contents

Acknowledgments

Many people contributed to the success of this trip and production of the report. Our thanks go to Eduardo Forno for coordinating extensive pre-trip arrangements and anticipating numerous details. Ana Martinet de Mollinedo, Marina de Montaño, and Lourdes Larea of the CI Bolivia office cheerfully accommodated the team and provided logistical support. Marcelo Sommerstein generously offered use of the Madidi camp and support from camp personnel. Padre Nivardo graciously provided assistance with field transportation. María Marconi and Isabel Mercado of the Bolivian Conservation Data Center provided valuable review of the appendices. Dr.Sydney Anderson generously shared records from his files on Bolivian mammals. Stephen Nash prepared the map of the region, Leonor Greenidge helped prepare the manuscript, and John Carr and Tom Schulenberg provided valuable editorial assistance. Finally, we gratefully acknowledge the MacArthur Foundation for establishing the Rapid Assessment Program, and the Beneficia Foundation and CI members for additional financial support.

Editors' Note

This report is both a synthesis and compilation of the findings of the RAP team. The Overview section provides highlights of the trip, and integrates suggestions and conclusions that were shared by team members after analyzing their results. The Technical Report provides more detailed information on species and natural communities, and is derived from separate reports submitted by the individual team members. Authorship of separate sections is attributed to the contributors whenever possible, though some sections were combined in order to avoid overlap or repetition.

Participants

SCIENTIFIC PERSONNEL

Theodore A. Parker, III

Ornithologist
Conservation International

Robin B. Foster

Plant Ecologist
Conservation International

Louise H. Emmons

Mammalogist
Conservation International

Alwyn H. Gentry

Botanist
Conservation International

Stephan Beck

Botanist
Herbario Nacional de Bolivia

Silvia Estenssoro

Botanist
Centro de Datos para la Conservación de Bolivia

Flavio Hinojosa

Mammalogist
Instituto de Ecología

FIELD ASSISTANCE

Hermes Justiniano

Pilot; Director, Fundación Amigos de la Naturaleza, Santa Cruz
Bolivia

Abel Castillo

Fundación Amigos de la Naturaleza

Brent Bailey

Director of Biological Programs
Conservation International

Edward Wolf

Editor/writer
Conservation International

EDITORS

Theodore A. Parker, III

Brent Bailey

Organizational Profiles

Conservation International (CI)

Conservation International (CI) is an international, nonprofit organization based in Washington, D.C., whose mission is to conserve biological diversity and the ecological processes that support life on earth. CI employs a strategy of "ecosystem conservation" which seeks to integrate biological conservation with economic development for local populations. CI's activities focus on developing scientific understanding, practicing ecosystem management, stimulating conservation-based development, and assisting with policy design.

Conservation International (CI)
1015 18th St. NW
Washington, D.C. 20036 U.S.A.
202-429-5660

CI–Bolivia
Avenida Villazón #1958, Of. 10-A
Casilla 5633
La Paz, Bolivia
(5912) 341230

Centro de Datos para la Conservación de Bolivia (CDC)

The Centro de Datos para la Conservación de Bolivia (CDC) is a private, nonprofit Bolivian organization created to contribute to the conservation of living resources of Bolivia. Viewing conservation as the appropriate use of these resources, with the goal of improving the quality of life of present and future generations, the CDC promotes sustainable development that integrates ecological, social, and economic factors with basic principles of conservation. The CDC's central mission is to provide the technical base for the development of strategies, policies, programs, and projects of protection and rational use of the country's biological heritage, and focuses on bridging the gap between those who generate biological information and those who use it.

Centro de Datos para la Conservación de Bolivia (CDC)
Calle 26, Cota-Cota Casilla 11250
La Paz, Bolivia
(5912) 797399

Herbario Nacional de Bolivia

Herbario Nacional de Bolivia is a botanical research center, established in 1983 by agreement between the Universidad Mayor de San Andrés and the Academia Nacional de Ciencias de Bolivia under the sponsorship of the Neotropical Flora Organization in La Paz.

The Herbarium's principal goal is the study of the flora of Bolivia, through floristic inventories, establishment of botanical collections, and development of basic and applied research projects, thereby contributing to scientific training at national and international levels in the Neotropical region.

Herbario Nacional de Bolivia
Casilla 10077
La Paz, Bolivia
(5912) 792582 and 792416

Instituto de Ecología de Bolivia

The Instituto de Ecología de Bolivia is a scientific training and research center of the Universidad Mayor de San Andrés de La Paz. Its principal objective is the enhancement of scientific capacity to resolve ecological problems in Bolivia. To realize this goal, the Instituto trains professional biologists with an ecological orientation and undertakes basic and applied research in programs of conservation, agroecology, and inventories of Bolivian flora and fauna.

Instituto de Ecología de Bolivia
Campus Universitario
Cota-Cota, Calle 27
Casilla Correo Central 10077
La Paz, Bolivia
(5912) 792582 or 792416

Fundación Amigos de la Naturaleza

The Fundación Amigos de la Naturaleza **(FAN)** is a private, nonprofit organization established in 1988. FAN's mission is to protect Bolivia's biological diversity. In collaboration with other nongovernmental organizations, the Bolivian government, and the international conservation community, FAN provides vital technical and financial assistance to Noel Kempff and Amboro National Parks and the Ríos Blanco y Negro Wildlife Reserve, protected areas comprising a total of almost 7.5 million acres. A long-term goal of FAN is to expand the system of protected areas from the current 2 percent to at least 10 percent of Bolivia's national territory through the establishment of new areas, training of conservation professionals, and a public environmental education program.

Fundación Amigos de la Naturaleza (FAN)
Av. Irala 421
P.O. Box 2241
Santa Cruz, Bolivia
(591-33) 33806
(591-33) 41327 (fax)

Overview

From 18 May to 15 June 1990, Conservation International's Rapid Assessment Program team (Louise H. Emmons, Robin B. Foster, Alwyn H. Gentry, and Theodore A. Parker, III) and counterparts from Bolivian institutions (Stephan Beck, Silvia Estenssoro, and Flavio Hinojosa) undertook rapid evaluations of fauna and flora of lowland and montane forests in the department of La Paz, on the eastern slope of the Andes in northwestern Bolivia.

The purpose of the expedition was to assess quickly the biological importance of a vast, largely unexplored wilderness area in Provincia Iturralde, along the upper reaches of the Ríos Heath, Madidi, and Tuichi. The region encompasses nearly 50,000 sq km of pristine forest and grassland, none of which currently receives protection under Bolivian law.

CONCLUSIONS

Results for this area of Bolivia indicate a high diversity of flora and fauna that rivals the richest known sites on the globe. Habitat heterogeneity, the general species richness of Amazonian and Andean forests and their proximity to each other, relatively high precipitation, and nearly complete absence of long-term human perturbation are among the related probable causes for the high levels of species diversity encountered by the group.

The region of northern La Paz, from the high Andes to the mouth of the Río Heath in the lowlands, is likely to harbor more bird and mammal species than any other comparable area of Bolivia. It is possible that more than 1,000 species of birds, or an amazing 11 percent of all bird species on earth, will eventually be recorded along a transect from the Andean grasslands near Lake Titicaca to the lowland forests and savannas near the mouth of the Río Heath. The region hosts what are probably the most

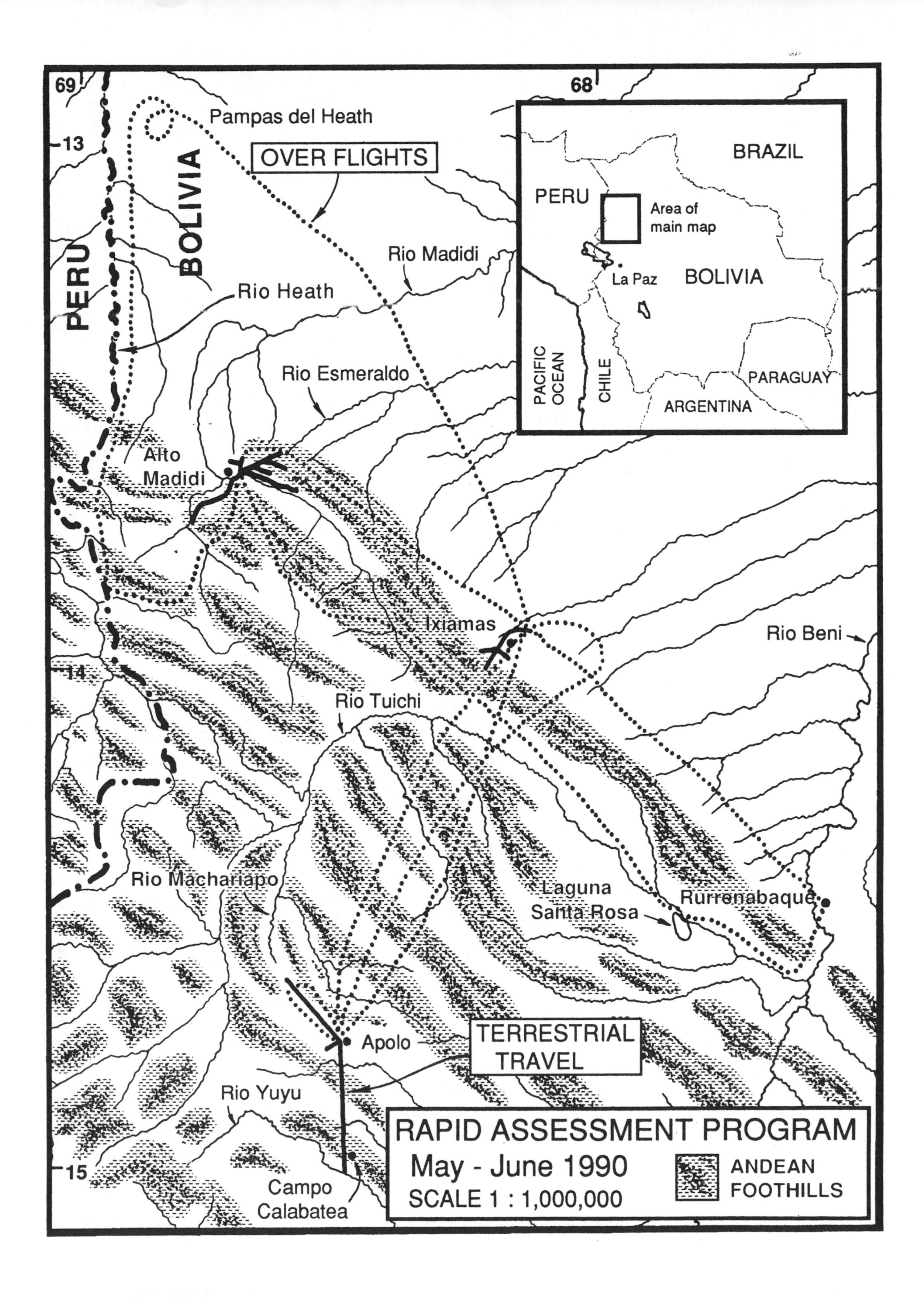
69
68
13
14
15
Pampas del Heath
OVER FLIGHTS
PERU
BOLIVIA
Rio Madidi
Rio Heath
Rio Esmeraldo
Alto Madidi
Ixiamas
Rio Beni
Rio Tuichi
Rio Machariapo
Laguna Santa Rosa
Rurrenabaque
Apolo
TERRESTRIAL TRAVEL
Rio Yuyu
Campo Calabatea
BRAZIL
PERU
Area of main map
La Paz
BOLIVIA
PACIFIC OCEAN
CHILE
PARAGUAY
ARGENTINA
RAPID ASSESSMENT PROGRAM
May - June 1990
SCALE 1 : 1,000,000
ANDEAN FOOTHILLS

Results for this area of Bolivia indicate a high diversity of flora and fauna that rivals the richest known sites on the globe.

species-rich forests in all of Bolivia on the ancient river terraces and the adjacent slopes of hills and ridges. Combined with the adjacent Tambopata-Candamo reserve in Peru, this area could become a biodiversity reserve unsurpassed in all of South America.

We saw in our overflights that most of a vast region of northern La Paz is uninhabited or virtually so. It is therefore in an ideal state for long-term land-use planning for conservation and sustainable development.

We had already suspected that the forests near Bolivia's frontier with Peru at the base of the Andes (known as the 'lower yungas') would be the richest in plant species of any forest in Bolivia. Our rapid assessment confirms this to our satisfaction. The region has good representation of the southwest Amazonian biota. One component, the floodplain forests, may be richer away from the Andes where the rivers meander more, and we suspect (but still do not know) that the middle or upper yungas (mountain slopes) are also richest in species near the Peruvian border.

The Alto Madidi and portions of the lower Madidi have virtually intact faunal and floral assemblages. Only in the immediate vicinity of the airstrip and camp have some of the *Cedrelinga* ('*mara macho*' in Spanish) trees been cut out. It is astonishing to see large *Cedrela* (*cedro*) near the lumber camp. The absence of *Swietenia* (mahogany, known as *mara* or *caoba*), either as a result of much earlier logging or for lack of suitable habitat, is very fortunate. It means there will be little logging pressure on this area for several more years. The absence of hunting has left even the large game animals near the rivers.

The Andes provide numerous habitat types, including four elevational zones above the lowlands, all of which host different bird and small mammal faunas. Our fieldwork at Alto Madidi (14 days) reveals how little we know of bird distribution in northern Bolivia (and birds are the best-known vertebrate group!).

The lowland savannas (pampas) are not *a priori* expected to be richer in plant species near the northwestern frontier with Peru, but they may contain more endemic species and certainly are less disturbed here than anywhere else north of thè Río Mamoré. Although its mammal list may not be large, the Pampas del Heath may be one of the only undisturbed natural habitats of its type, and serves as a refuge for species that are persecuted elsewhere and in need of protection. The pampas clearly represent an important conservation priority in Bolivia and South America.

At Ixiamas we found several grassland bird species that are declining throughout most of their ranges in central South America (e.g., Cock-tailed Tyrant, *Alectrurus tricolor*, and Black-masked Finch, *Coryphaspiza melanotis*). That large populations of such species survive in northern and central Bolivia underscores the conservation opportunities that still exist in this region.

In addition to the northwestern border evaluated here, one could make a similar argument about the conservation importance of Bolivia's other border regions. The southern border with Argentina and Paraguay has probably the greatest richness within Bolivia of subtropical biota. The eastern border with Brazil (e.g., the Serranía de Huanchaca) certainly should have the greatest richness within Bolivia of Brazilian Shield species. The cerrado vegetation is disappearing in Brazil to a much greater degree than in the smaller area in Bolivia.

Bolivia is not at the heart of any one of these large biotas, but it can claim to be the most important transitional country between the major biotic regions of southern South America. Our goal is not to single out Bolivia as the ideal parkland for southern South American biota, but rather to draw attention to its enormous biological wealth and relatively low population pressure. The pressures of growing population and development on the neighboring countries may in many cases be too strong to protect these biotic systems. In Bolivia it is still very possible.

CONSERVATION OPPORTUNITIES

The biological surveys discussed in this report indicate that lowland and montane forests in northern La Paz support the richest plant and animal communities in the country. We therefore suggest the following opportunities for their protection:

1 **Establishment of a large conservation unit in the department of La Paz that would include large areas of lowland and foothill forest along the upper Rios Heath and Madidi (including the entire upper drainage of the Alto Madidi), and the higher ridges to the south (e.g., the Serranía de Tutumo).** The very high biodiversity of the Alto Madidi region, particularly that of lowland forests at the base of the mountains, but also of montane forests on outlying and higher ridges to the west, could be protected in a reserve that would border the Tambopata-Candamo Reserve lying along the Peruvian side of the Río Heath. The resulting bi-national reserve would encompass some of the richest forests in upper Amazonia, including the most diverse forests in all of Bolivia. Up to 12 percent of all bird species on earth, for example, and the highest plant species diversity yet reported from Bolivia, have been found between 400 and 4,000 meters in the headwater region between the Río Tambopata, Peru, and the Río Alto Madidi, Bolivia.

2 **Extension of the boundaries of this Alto Madidi protected area southward along the Bolivia-Peru border to include the full gradient of montane forests up to treeline (including the humid paramos).** The watershed value of these forests, in addition to their biological importance, is inestimable. The upper montane section of the proposed border reserve could be extended southeast to the Apolo-La Paz highway currently under construction. The benefits of limiting colonization along such roads (e.g., to limit erosion that results from deforestation for subsistence agriculture) are often overlooked or ignored.

3 **Establishment of a second conservation area for watershed management, the protection of diverse ecosystems, and development of ecotourism in the drainage of the Río Tuichi.** This could include the higher elevations of mid-montane forest south of the Tuichi up to the burned plateaus northeast of Apolo. This would encompass several long trails from Apolo to the lower Tuichi which could be maintained for trekking, patrolling, and research, now that new access roads and airplanes make the trails of little use to the population around Apolo.

The tall, lower montane forests of the Serranía de Eslabón, along the western side of this valley, probably support the richest montane plant and animal communities in Bolivia. The existing tourist lodge at Laguna Santa Rosa would be an excellent base for biological inventories of the surrounding lowland forests as well as of montane forests to the west. Like the Alto Madidi site described above, the lodge is strategically located for monitoring economic development along the river.

4 **Creation of extractive reserves around the proposed Madidi conservation unit could lead to the rational, long-term exploitation of forest-based products.** Economically important forest resources, including large populations of valuable trees such as *Swietenia macrophylla* and *Cedrela odorata*, occur in this region. Unfortunately, these are being exploited rapidly, and perhaps not rationally, by a number of logging companies that have already

It is possible that . . . 11 percent of all bird species on earth, will eventually be recorded along a transect from the Andean grasslands near Lake Titicaca to the lowland forests and savannas near the mouth of the Río Heath.

been granted concessions. Establishment of extractive reserves where harvesting of such species could be monitored and studied would benefit the Bolivian economy far into the future. Forests along the lower Río Heath, support large numbers of castaña trees (*Bertholettia excelsa*), which already yield economic rewards for local people.

---

5 Construction of a dual-purpose biological station and guard post at the site of the recently abandoned Aserradero Moira sawmill at Alto Madidi. The strategic importance of this base for scientific studies as well as for surveillance cannot be overstated. Access by airplane is a wonderful convenience for the operation of a station, requiring only maintenance of the airstrip. Also essential is maintenance of the road to Ixiamas. An improved network of trails around the station would facilitate biological explorations of the surrounding forests. Floral and faunal inventories of the unexplored ridges, some as high as 1,800 m, to the west of Alto Madidi would reveal the presence of large numbers of endemic species. As a guard post, the station could also monitor exploitative practices such as timber extraction and gold mining. At present there seems to be little control over such activities.

---

6 Establishment of a protected area encompassing the Bolivian Pampas del Heath and surrounding forests, the only extensive areas of undisturbed pantanal-like grassland remaining in northern Bolivia. This could be accomplished without much difficulty, as the human population along the lower Río Heath is very small and few if any cattle graze there. A biological reserve would protect healthy populations of many large mammal species, as well as a high diversity of birds (see Appendix 2). Pampas on the Peruvian side of the river are already protected by law as part of the Tambopata-Candamo Reserve but are very small compared to those on the Bolivian side. Bolivians and Peruvians living along the river below the grasslands could find opportunities for employment in managing and protecting the reserve.

As one of the last remaining Bolivian pampas not overrun by cattle and seemingly unique in floristic composition, the Pampas del Heath should be kept free of roads. The river and small airstrips provide adequate access and control.

---

7 Protection of representative examples of the unique plant communities that occur in the semiarid valleys of the northern yungas, such as the Río Machariapo dry forest. Numerous potentially threatened plant and animal species may be restricted to small geographic and elevational zones within the yungas region. There is an urgent need for rapid and intensive biological inventories of such areas.

---

8 Promotion of reforestation projects in the densely settled inter-Andean valleys such as that surrounding Apolo. We were surprised to learn that soldiers at the local military base are sent up to 50 km away to gather firewood and building materials from existing native forests. Eucalyptus plantations, which commonly supply firewood to the local populations in other dry Andean valleys, have not been cultivated here. We would encourage the planting of other tree species, preferably fast-growing native species.

---

9 Extension of a reserve or protected area along the Peruvian border to the northwest tip of Pando. This would add many primate and small mammal species not found further south, and would result in a reserve rivalling Parque Nacional

Manu of Peru as the biologically richest protected area in the world.

SUMMARY OF FIELD WORK

Over a two-week period, the group surveyed lowland evergreen forests at Alto Madidi, a logging camp on the west bank of the Río Madidi about 20 km south of the Peruvian border and 100 km northwest of the nearest small town, Ixiamas. Another ten days were spent in the area of Apolo, 125 km southwest of Ixiamas, in mid-elevation wet and dry forests. Brief field time was spent on the low ridges and savannas around Ixiamas. Overflights crossed the broad Pampas del Heath northwest of Ixiamas along the Bolivia-Peru border, and southeast of Ixiamas to Rurrenabaque on the Río Beni.

Field methods varied according to each specialist. Gentry and Estenssoro collected plants and data on woody species along a series of transects 2 meters and 50 m long in different types of forests. Foster made qualitative assessments of the vegetation structure and plant community composition of all habitats within walking distance, made lists of all the plant species observed, and contributed voucher specimens of important plant species not found in the transect. During overflights, he identified forest types in the region from the air. Beck did general collecting of plant specimens, and obtained specimens of grass species for his own research. Emmons made daily and nightly forest walks (totalling 85 hours at Alto Madidi), recording all mammal species seen and heard, in addition to those caught in small live traps. Hinojosa mist-netted bats and also trapped small mammals. Parker and Castillo surveyed birds with the use of tape-recorders and mist nets.

At the Alto Madidi camp, where weather, logistics, and absence of human perturbation were optimal, Parker identified 403 bird species in 14 days along a transect of roughly six kilometers by 200 meters. Of these, nine were first records for the country, 30 were recorded in the country for only the second time, and 52 species were new for the department of La Paz. Among the 45 species of mammals Emmons and Hinojosa identified in that period were abundant populations of tapirs and spider monkeys, testimony to the absence of local human impact. Of particular interest were two species of mammals previously unrecorded for Bolivia: a little big-eared bat (*Micronycteris nicefori*), and a spiny tree rat (*Mesomys hispidus*). In a rare sighting of the Short-eared Dog (*Atelocynus microtis*), the individual carried a frog in its mouth; this is the first record of food habits for a wild individual.

Botanical results from the Madidi camp are equally exciting, showing an unusually high diversity of plant species. Forests on floodplains, high terraces and slopes, and ridgetops, each host distinctive floras and contribute to the overall richness of the area. In a tenth-hectare sample on forested low rolling hills, Gentry found 204 species greater than 2.5 centimeters in diameter. According to his analysis, the average moist forest transect yields 152 species. Many of the species found at Madidi are new to Bolivia; some of them are likely to be new to science.

While we were fortunate to have two productive weeks in the lowland forest, weather and logistical problems limited extensive coverage of the region's savannas. Our brief exposure to Pampas del Heath and the area around Ixiamas, as well as our past experience on the adjacent Peruvian pampas, however, clearly point to the grasslands of the region as a high priority for further exploration and conservation work. Botanically, initial indications are that the region's various savannas constitute a far more complicated mosaic of isolated habitat islands than a glance at the map would suggest.

Future rapid assessments are urgently needed in Bolivia concentrating on specific habitats, physiographic types, and vegetation types . . .the pampas and dry valleys must be an urgent priority because so few remain untouched by human activity.

Remarkable as these findings are, they represent only the first step in a recommended research and conservation endeavor for the region. Much remains to be discovered: A rapid assessment is a preliminary indicator of an area's biological importance and appropriately generates more questions than answers. From this expedition, however, it is clear that longer-term, systematic biological inventories will confirm the value of northern La Paz as a repository of impressive biological diversity that is of global significance. Simultaneously, information from this initial assessment is sufficient to call attention to the conservation value of the region and to stimulate initiatives for its long-term management and protection.

DIRECTIONS FOR FURTHER RESEARCH

Rapid Assessments

The RAP team's overview of the mosaic of vegetation types of northern La Paz serves its purpose by focusing international attention on the importance of this region. Future rapid assessments are urgently needed in Bolivia concentrating on specific habitats, physiographic types, and vegetation types.

The pampas offer a good starting point. It became obvious to us from our brief surveys of the westernmost pampas of northern La Paz that they differ radically from each other as well as differing from pampas in the Beni and Santa Cruz region. To evaluate the conservation importance of the different Bolivian savannas, an expert team of botanists, zoologists, and ecologists should be organized soon to inventory and compare them.

The team should go from one to another in succession, surveying all the different habitats within each pampas area. A smaller team check should be done in high-water periods.

Floodplains should be evaluated for variation along the length of rivers, and comparisons made between floodplains from north to south and east to west. Alluvial fans should be compared along the whole length of the base of the Andean foothills. From this, scientists could determine if the Beni river has really been a major isolating mechanism, blocking the migration of species from northern Bolivia to central Bolivia as some evidence suggests. Similarly, the low foothill ridges should be compared along the base of the Andes. This would be especially instructive if different kinds of ridges could be categorized as they have been in this report.

Up to now, collecting in montane forests has extended to little more than along the few good roads down from the altiplano. A much broader comparison is needed to distinguish different communities if they exist, to determine the variation in altitudinal differences of these communities, and to discover important differences between areas in diversity and endemism. It is urgent that the inter-Andean dry valleys be studied to find how many, if any, are still reasonably intact and how much variation exists among them.

All of these suggested surveys are needed soon. But the pampas and dry valleys must be an urgent priority because so few remain untouched by human activity. Many other important habitats or vegetation types in Bolivia—ranging from puna to chaco woodland—require an initial rapid assessment.

Remote Imagery

Many regions of special interest to conservation are subject to frequent cloud cover or have not been given priority by government officials for remote-sensing information. Frequently, modern radar and multispectral images are not immediately available to the RAP team. Nevertheless, once the image information is available, and especially once it can be processed by computer to reveal fine-scale patterns in ground cover, the information can quickly be put to use and extrapolated to larger areas for mapping. It remains to be seen how many of the wet-forest plant communities recognizable on the ground can be distinguished with these images. It is already clear from this trip that satellite images are not sufficient (at least by eye) to distinguish several of the important plant communities.

Geology and Climate

Most of the plant communities are strongly subject to the effects of geological substrate and climatic variation. Scientists could probably map community distribution without ever looking at an organism if enough information were available on the geographic distribution of geology and climate. Unfortunately, the geological maps that exist are crude and inappropriate, focusing on fault lines and the age of the substrate. Within any geological age there can be a huge array of different kinds of rock, often with radically different effects on the vegetation (e.g., quartz sandstones and limestones). But we have no maps indicating the kind of rock exposed on the surface nor at the level where plants have their roots, and we lack an analysis of these substrates for the characteristics that are important to the soil and organisms above them. Most soil maps based on soil samples fail to take into consideration the effect of the vegetation itself on the soil, and the importance of deeper layers that are reached by plant taproots.

Similarly, climatic maps rarely provide information of importance to the organism. Total rainfall, for example, is not as important as its frequency during the year, its variation from year to year, its extremes, the cloud cover and wind over an area, the amount of fog precipitation, and draining or flooding once the precipitation reaches the ground.

More research and mapping of this nature, none of it very easy, will tell us much of what we need to know about the distribution and maintenance of biotic communities important to conservation. Short of that, the plant communities themselves can be the best indicators.

Follow-up Inventory and Field Guide Production

The Madidi-Tuichi region, with its considerable importance to conservation in Bolivia, is clearly a priority for thorough inventory. Field guides developed to identify the different groups of organisms and describe the communities in this area will almost certainly cover the great majority of species in the rest of the lower yungas in Bolivia, as well as the southernmost populations of many species known mainly in countries to the north—but with a better chance of survival in Bolivia. Any follow-up in providing such reference tools will have an impact far beyond the Madidi area itself.

Technical Report

ALTO MADIDI REGION

From 18-31 May 1990, the RAP team surveyed lowland evergreen forests at Alto Madidi, a lumber camp on the south bank of the Río Madidi about 20 kilometers south of the Peruvian border and 100 kilometers northwest of the nearest small town, Ixiamas. This camp was a perfect base for study, being situated in lowland forest at the base of the Serranía del Tigre, the easternmost ridge of the Andes. The camp was within walking distance of a variety of forest types, including young river-edge forest, more mature floodplain forest, and older forest on ancient river terraces and slopes of hills and higher ridges to the south and west.

Physiography of Alto Madidi, Bajo Tuichi, and the Foothill Ridges (R. Foster)

Ridges

According to the available geological maps, the foothill ridges that reach about 1,000 m altitude are composed of Ordovician, Devonian, and Cretaceous rock layers pushing up through undulating hills of Tertiary age. The older rock is of the same age as that forming the higher (up to 2,500 m altitude) mid-elevation ridges of the middle yungas. These are separated from the foothill ridges by a broad Tertiary trough known as the Madidi-Quiquibey Sincline, along which pass the Alto Madidi, lower Tuichi, and Quiquibey rivers. The Río Beni bisects this trough between Rurrenabaque and the Serranía Chepite.

The foothill ridges are mostly very steep on both flanks, forming knife-edge crests. Most higher ridges exposed by landslide have a pale whitish or yellowish color in contrast to the red of the Tertiary hills. But the rock is far from uniform in composition. Large portions of the main ridge and smaller outlying ridges are composed of a quartzite-like material often revealed in rectangular blocks on the tops of ridges, and with considerable eroded sandy sediment around the base. Recent landslides may account for somewhere between a fifth to a tenth of the surface area.

Hills

The low Tertiary hills and terraces are mostly composed of a distinct dark pinkish-red clayey rock. On exposed banks of the high, non-flooded Quaternary terraces, a few meters of red Tertiary strata are always visible below the upper layer of gravel and boulder sediment. The rivers are apparently wearing down through the recent sediment and into the older material while simultaneously the Tertiary strata are being lifted up by the same tectonic forces that are raising the foothill ridges. Anomalous within the red hills are a few erosion-resistant shields of what appear to be sandstone or quartzite (e.g., the Serranía del Tutumo southwest of Ixiamas).

Where the Tertiary strata are compressed and subject to faulting, they are raised in blocks at a steep angle forming irregular hills. These blocks of soft red material are especially prone to landslide. Throughout the area of Tertiary hills (e.g., those along the east side of the lower Tuichi), an impressive one-third of the surface area consists of exposed recent landslides or young successional vegetation on landslides. Though this could have been caused by severe earthquake, it is more likely the consequence of the angle and softness of the strata and is probably a permanent dynamic condition.

Alluvium

The hills are interrupted by landslide alluvium from higher ridges and reworked sediment from the current river systems. The sediment is of course derived from both the high ridges as well as the hills, and tends to sort out or mix together differently depending on the conditions of deposition. This results in a mosaic of soils, at least in the recent alluvium, the most obvious difference being between sandy and clayey soils. The abundance of sandy beaches on the Madidi is a testament to the importance of quartzites or sandstones in the surrounding ridges. The lack of meander formation in the rivers draining the Madidi-Quiquibey Sincline is attributable to both the slope of the drainage and to its confinement by ridges and hills.

The river beaches in valleys with only weakly developed meanders are nearly static. They flood many times a year during the rainy season, which eliminates the temporary vegetation, but they only rarely or slowly form levees from which a permanent vegetation can develop. More frequently, the river changes course abruptly, forming an island, leaving a low abandoned channel, or leaving a broad pile of landslide rubble.

To the northeast of the foothills, the erosional sediment abruptly spreads out into a series of overlapping alluvial fans formed by landslides and by all of the streams draining from and through the foothills. These fans of well-drained alluvium form a band of variable width along the base of the foothills, interrupted by inundated or poorly drained areas with increasing frequency at greater distance from the foothill ridge. As is typical on alluvial fans, the streams draining the ridge constantly make major jumps in their courses. They are appropriately referred to as arroyos since they are frequently dry, having only small rainfall catchment areas and considerable underground flow in the loose rocky soil.

Throughout the hills . . . along the east side of the lower Tuichi, one-third of the surface area consists of exposed recent landslides or young vegetation on landslides . . . it is probably a permanent dynamic condition.

Plant Communities of Alto Madidi, Bajo Tuichi, and the Foothill Ridges (R. Foster)

Beach and early riverine succession

The annual beach community of herbs and seedlings of woody plants only flourishes toward the end of the dry season. We were not able to determine if this was a particularly rich or poor community based on the tiny plants starting to appear on the banks of the Alto Madidi in late May. Nearly all of them would be weedy species of minimal conservation interest because of their pre-adaptation for colonizing human clearings.

The successional flora of river deposits along the upper Río Madidi is typical of most of upper Amazonia, starting with the fast-growing treelets *Tessaria integrifolia, Baccharis salicifolia, Salix humboldtiana,* and *Gynerium sagittatum.* In the later stages there is a predominance of such species as balsa (*Ochroma pyramidale*) that succeed better on the sandy alluvium predominating on these rivers. *Cecropia membranacea,* more abundant on mud or silt beaches, is present but not common. Some species absent from the Alto Madidi study area may in part be limited by the paucity of finer silt deposits. In any case, such forests are neither as abundant or as species-rich as one would expect to find on a more-meandering, silt-depositing river system. From the overflight it appeared that the lower Madidi would be such a system.

What distinguishes the area from the average young floodplain forest on meandering rivers is the absence of strong dominance by the canopy species *Ficus insipida* and *Cedrela odorata*, and a greater representation of the smaller interstitial species such as *Acacia loretensis, Nectandra reticulata, Terminalia oblonga*, and species of *Inga, Erythrina*, and *Sapium.* This is to say that the community had much more "evenness" in relative abundance among the component species. Many of the scattered successional forests on the Alto Madidi floodplain may result from the sudden destruction and deposition following massive landslides in the headwaters or radical shifts in river course. The relative inability of the fig and cedro to colonize in abundance may be somehow related to dispersal and establishment problems on such substrate.

On the upper pebbly and rocky beaches a more stable community develops, consisting usually of *Imperata* grass, the shrubs *Calliandra angustifolia* and *Adenaria floribunda*, and on steeper banks, the tree *Pithecellobium longifolium.* Thickets of bamboo (*Guadua* sp.) occur locally on the floodplain; they are associated with areas of forest disintegration where floodwaters spill over a bank with sufficient force to take down trees or deposit a smothering layer of sediment. The dense, spiny stands seem capable of persisting in one spot for many years, probably until the population finally flowers or gets swept away by the river. These thickets on low floodplains could have been caused by human clearing but we found no evidence for this to be so. However, the bamboo thickets on higher floodplain terraces did seem to be from human intervention, the only source of forest disturbance large enough for bamboo to become established within this community.

Older floodplain forest

The more mature floodplain forest, higher but still subject to occasional flooding, is similarly not as abundant or as rich in species as on more extensive floodplains to the north away from the mountains. It would, however, be misleading to imply that it is impoverished, since there are hundreds of woody plant species, and it is significantly richer than floodplain forests seen in the Beni to the south (Foster, unpubl.). Nevertheless, this forest is mostly dominated by a single species, *Poulsenia armata,* among the canopy trees, with occasional individuals of the expected large *Dipteryx, Hura, Ceiba pentandra, Brosimum alicastrum,* and *Sloanea cf. obtusifolia.* In the understory *Astrocaryum macrocalyx* predominates, but with conspicuous numbers of *Socratea exorrhiza, Iriartea deltoidea, Otoba parvifolia, Iryanthera jurense, Trichilia pleeana*, and *Quararibea wittii*.

High terraces and low ridge slopes

Probably the most species-rich forests in all of Bolivia are on the ancient river terraces and especially the slopes of the hills and ridges. In most areas these would be separate physiographic entities with sufficiently distinct floras to be treated as distinct communities. Here their close proximity and overlap in flora justify lumping them together.

The terraces are more uniform in topography than the slopes and as a consequence are also more uniform in community composition throughout, and apparently less species-rich. Virtually all the species of the terraces were also found on the slopes, but the reverse was far from true. The terraces are probably more diverse than they would otherwise be because of the seed rain coming in from the slopes.

The slopes, in addition to frequently having a year-round source of groundwater from higher up, have more heterogeneity in moisture conditions, soil conditions, light conditions, and a patchwork disturbance regime in the form of landslides and lateral slumps along streams. The latter not only permits the long-term survival of numerous rare species from chance colonization of the clearings, but it also provides the conditions (long-term minimal competition) for many of the well-dispersed floodplain species to get established. An example is *Poulsenia*, which is found down on the floodplain and up on the slopes, but not on the intervening high terraces.

These forests have unusually large adult populations of *Ficus sphenophylla*. These enormous-crowned, large-fruited trees seem to occur at a density of one per two or three hectares throughout the area. While not registering as abundant along transects by counting the number of tree stems greater than 10 cm in diameter, they make up a significant fraction of the biomass and crown area, and probably of edible fruit production. Many of the tapir scats encountered were made up principally of fruit material of this species, and *Ficus sphenophylla* fruits are likely an important food resource for many other vertebrates and insects, especially as their fruiting is not confined to one season.

Other conspicuously abundant large trees include *Apuleia, Cedrelinga, Copaifera, Huberodendron, Hyeronima alchorneoides, Manilkara, Parinari, Pterygota, Sterculia, Tachigali*, and *Tetragastris*. Abundant smaller trees are *Apeiba membranacea, Batocarpus amazonicus, Inga*, various Lauraceae, *Pourouma minor, Tetrorchidium*, and *Virola calophylla*. Abundant treelets and shrubs are *Capparis sola, Coussarea* sp., *Hirtella racemosa, Hyospathe elegans, Palicourea punicea, Pausandra trianae, Perebea humilis, Piper augustum, Piper obliqum, Pleurothyrium krukovii, Siparuna decipiens*, and *Stylogyne cauliflora*.

On the alluvial fans of the larger drainage canyons at the base of the main foothill ridges (e.g., near Ixiamas), the distinction blurs between floodplain, terrace, and slope communities. The species distinctive of each are often found mixed together, sometimes even with ridgetop species. Species distributions on these areas tend to be extremely patchy, often with local dominance by just a few species even when the species-richness of the whole area is high.

(A. Gentry): Structurally, the Madidi lowland forest is fairly typical of Amazonian forests. The most unusual structural feature is the relatively high density of lianas (93 lianas 2.5 cm diam. in 0.1 ha vs. a Neotropical average of 69). Also noteworthy is that many of the climbers are hemiepiphytic, with the fern *Polybotrya* (the second most common species in the entire sample) especially prevalent. This is a feature usually more closely associated with cloud forests than with Amazonian forests. The implication is that the Madidi area enjoys unusually high rainfall. The density of trees 10 cm dbh is also greater than normal in Amazonia. Many of the larger trees are palms, with seven individuals of *Iriartea* and six of *Euterpe* making these the two most common taxa of trees 10 cm dbh. This structural prevalence of palms contrasts with their relatively low diversity.

Wet ridgetops

We were unable to visit any of the highest foothill ridgetops which approach 1,000 m in altitude. On a somewhat lower (600-700 m) ridgetop at the northernmost tip of the outlying range near the Alto Madidi (a ridge summit touched by clouds intermittently, though with few large trees), the woody plant composition was almost indistinguishable from that of the lower slopes. The most striking difference is the abundance of epiphytes and the moss cover on trunks and branches on the ridgetop. Very few species of trees from mid-elevation wet forest were found (*Clusia*, *Calyptranthes*, and another Myrtaceae).

The higher ridges probably have the same species composition, but with an increasing frequency of mid-elevation wet forest species and an increasing density and diversity of epiphytes.

Dry ridgetops

On the steep slopes below the wet ridges and on lower ridgetops that do not reach the cloud level, there is a distinctive community of plants. Some of the species (e.g., *Rinorea viridifolia*, *Mouriri myrtilloides*) occur in the lower areas but have their peak abundance on the dry ridges. Many others seem to be found only in this habitat. The most characteristic plants are a new tree species in the Sterculiaceae that was recently described by Gentry as *Reevesia smithii* and previously collected on the foothills to the south in Beni, and several shrub species mainly in three families: Rutaceae (*Erythrochiton*, *Galipea*, *Esenbeckia*, etc.), Euphorbiaceae (e.g., *Acidoton*), and Violaceae (*Rinorea* spp.).

My suggestion as to why this community is distinct is that it supports species with unusual drought tolerance. The drainage, wind exposure, and the annual lack of precipitation during several months of dry season probably impose severe water stress on seedlings as well as adult plants, and only the most tolerant plants survive. Many of these plants are apparently unable to compete in the more mesic conditions elsewhere. The only other feasible explanation is that these ridges have some unusual geochemical composition. However, judging from other places where I have seen these same species growing, the first hypothesis seems the more probable.

(A. Gentry): The forest on the low ridgetops that form the first row of Andean foothills is distinctly different. A 0.1 ha transect sample between 360 and 380 m on the first ridge was made and compared with the transect on the lower slopes. Incredibly, there is an overlap of only about 23 spp. out of the approximately 200 spp. in each of these samples! (Identifications in *Inga* and Lauraceae could modify this number slightly). In well-known and speciose families like Bignoniaceae and Palmae, none of the species occur in both habitats.

(A. Gentry): The ridgetop forest differs from the lowland forest not only at the species level, but also in the relative importance of different families. Although Leguminosae (18 and 20 spp.) is the most diverse family in both samples (as nearly everywhere in Amazonia), the next most speciose family on the ridgetop is Rubiaceae (ca. 18 spp. vs. 3 in the lowland sample). This remarkable diversity is more characteristic of premontane than of lowland forests. Other speciose families noticeably better represented on the ridgetop are Euphorbiaceae, Bombacaceae, and Sapindaceae.

Several interesting montane elements found only on the ridge include *Aiphanes*, *Condaminea*, and *Styrax*.

Quartzite ridgetops

On the ridges or portions of ridges with quartzite outcrops, there is a high proportion of genera and species not seen anywhere else in the foothill region. The trees are short with small, tightly packed crowns of uniform height, giving the ridge a bald or slick look from a distance. Many of the taxa, such as *Graffenriedia*, *Aparisthmium*, *Freziera*, *Styrax*, *Asplenium rutaceum*, and *Maprounea*, are known to be associated with more acidic, nutrient-poor soils. The upper slopes of these ridges are dominated by two species of *Aspidosperma* and *Humiriastrum* trees, and the shrub layer is dominated by *Geonoma deversa* and an unusual, small-leaved *Psychotria*. Among the other distinctive plants throughout the knife-like ridges are the lianas *Distictella* and *Bredemeyera*.

From their altitude (500-700 m) and position with respect to the main foothill ridge, the quartzite ridges visited on this trip would otherwise have qualified as "dry ridgetops," though with very few species in common with the latter. Perhaps the substrate has different water retention characteristics than other ridges. In any case, it remains to be seen what would be the community composition of a "wet" quartzite ridge higher up in the clouds.

Phytogeography of Alto Madidi, Río Bajo Tuichi, and the Foothill Ridges (A. Gentry)

Floristically, there seem to be some minor peculiarities associated with the Madidi forests. Although the forest is composed of typical Amazonian elements, with Leguminosae and Moraceae being the most speciose families represented in the transect sample (followed by Bignoniaceae, Lauraceae, Sapotaceae, Melastomataceae, Meliaceae, Myristicaceae, Myrtaceae, and Chrysobalanaceae); some intrinsically tropical families such as palms, Annonaceae, Connaraceae, Lecythidaceae, and Piperaceae are relatively underrepresented compared to adjacent Peru. Others like Sapotaceae, Meliaceae, Euphorbiaceae, and such southern families as Myrtaceae and Proteaceae, are unusually well represented. To what extent these may be sampling artifacts, perhaps associated with edaphic peculiarities of the limited area surveyed, or to what extent they represent broader biogeographical patterns remains unclear. For several families like Rutaceae (3 spp.), Moraceae (18 spp.), and Bignoniaceae (16 spp.), the Madidi lowland forest 0.1 ha sample is among the most species-rich yet sampled in the world. The unusually well-represented families are generally those associated with relatively rich soils, and it is likely that the Madidi soils are relatively fertile for Amazonia, which may also account for the high mammal biomass of the area.

Some of the plant species are rare ones, many are new to Bolivia, and a few are new to science. A new *Arrabidaea* (Gentry 70382) will be described as *A. affinis*, a new *Distictis* (Gentry 70258) as *D. occidentalis*. Several other Bignoniaceae are new to the country. At the generic level, the palms *Wettinia* and *Wendlandiella* are also new to Bolivia, as are the genera *Anthodiscus* of the Caryocaraceae, and *Pterygota* of the Sterculiaceae, the latter being quite abundant here. Even many of the common species such as *Aspidosperma tambopatense* (the latter only recently described from Peru) have apparently never been collected in the country before.

Human Impact on Vegetation (A. Gentry and R. Foster)

Except within a kilometer or (at most) two from the airstrip and sawmill, the forest appears to be virtually undisturbed, unless ma-

The conservation implications are obvious: Protection of a large section of rainforest in this region will preserve as good a sample of biological real estate as would conserving a more equatorial one.

hogany (or 'mara,' *Swietenia macrophylla* in the Meliaceae) had been removed sometime earlier when it was a military prison camp. This area was remote to begin with and the mere presence of the prison may have considerably discouraged lumbering. Ironically, the location of the sawmill was based on the misidentification from the air of the numerous 'mara macho' (*Cedrelinga catenaeformis*, Leguminosae) which greatly resemble mahogany. Once constructed, the mill was mainly used to saw mahogany brought in from the Amazon plain to the east. The abundant remains of former prisoners and associated rumors of ghosts have also had a considerable effect on keeping the lumbermen from wandering or hunting far from the camp.

Cedrelinga is a very valuable and sought-after wood in the rest of the upper Amazon, and it is now greatly reduced in most of Latin America. But in this area, it has only been cut down right near the sawmill. There were several large intact trees right along the main entry road. Among the large trees in the transect sample is a *Cedrela* 52 cm in diameter. Apparently in this region of Bolivia, one of the last places where mature mahogany trees still occur in appreciable numbers, harvesting pressure on even the second most valuable hardwood species has been negligible to date, another argument for the conservation importance of the region.

Finally, the high density of large lianas (5 individuals greater than 10 cm in diameter in 0.1 ha) provides an independent indicator that the forest is old, since an abundance of large lianas is probably the single best physiognomic indicator of very old or primary forest.

Plant Diversity (R. Foster and A. Gentry)

(R. Foster): The unusually high plant species diversity of the Alto Madidi area results from the close juxtaposition of different floras, including those of the current floodplain on the flatlands and those of the adjacent foothill ridges. We made note of 113 families, 528 genera, and 988 species in a little over a week. (See Appendix 7.)

The floodplain flora alone at Alto Madidi is probably significantly less rich in species than floodplain forests further out on the Amazon plain, even along the same river. The ridges and slopes constrain any large development of floodplain forest; meanders are limited and oxbow lakes are rare. This confinement probably restricts dispersal into and maintenance of species on the active floodplain. The recent floodplain of the upper Río Madidi, while not poor in species, does not have the richness of other larger meandering rivers I have seen to the north in Peru.

However, the adjacent hills and higher ridges provide numerous small refuges and disturbance opportunities maintaining a large plant species pool that is available to colonize the nearby lower slopes and the non-inundated high floodplain terraces. They also provide a heterogeneous soil derived from the alluvial mixing of two or more distinct geological substrates. Consequently these areas have an especially high diversity of plant species in comparison to areas of extensive and uniform old floodplain terraces without adjacent hills, and are comparable in richness to almost any other upper Amazonian forest.

While neither the ridge flora nor the recent floodplain flora itself may be unusually rich in species in comparison to similar habitats farther north in Peru, the mixing of these floras, in the intermediate habitats which dominate this area probably accounts for part of the high regional diversity.

Except for on the ridgetops, the precipitation in this area does not appear to exceed 2 to 3 m per year. Even if the total rainfall is higher, the land is probably still subject to a prolonged annual dry season of 3 to 4 months. The epiphyte load on the trees appears little different from that along the Río Manu, Peru, which has an annual rainfall of less than 2 m. The higher ridges (above 900 m) clearly receive more precipitation, at least in the form of fog drip from the frequent clouds; mosses, leafy liverworts, and filmy ferns are abundant. Lower slopes benefit from this higher precipitation on the ridgetops through a year-round abundance of groundwater percolating down from above. The local habitats most subject to drought are probably the lower ridges and higher hills that do not reach cloud line, and secondly, the high, flat terraces on sandy-gravelly alluvium. These areas probably suffer a severe drop in water table when rainfall declines during the dry season, though the latter habitat may get some moisture from early morning low fog.

(A. Gentry): The most striking aspect of the Río Madidi forest is its high diversity. The 204 species greater than 2.5 cm dbh in a tenth hectare transect-sample of the forest on the low rolling hills just behind the floodplain is as high as in the Iquitos, Peru area. It is significantly higher than the values for equivalent samples in adjacent Madre de Dios, Peru, which average about 140 species. In general, forests from areas with stronger dry seasons are less diverse, and in Amazonian Peru the more southerly forests are the least diverse. One could therefore assume that the Bolivian forests would be (relatively) floristically depauperate. Thus the extremely high (plant community) diversity of the Alto Madidi area is a distinct surprise. Although actual data are scarce, the high plant diversity of the Madidi area may be related to the unusually high precipitation at the base of the adjacent "corner" of the Andean foothills. The conservation implications are obvious: Protection of a large section of rainforest in this region will preserve as good a sample of biological real estate as would conserving a more equatorial one. Data from plant transects corroborate the high diversity noted in the area (See Table 1.)

Birds of Alto Madidi (T. Parker)

The avifauna of the lowland forests in the Alto Madidi area was found to be unusually rich and is similar to those of two well-studied localities in nearby southern Peru, the Tambopata Reserve (Parker 1982; Donahue and Parker, unpubl. data) and the Cocha Cashu Biological Station in Manu National Park (Terborgh et al., 1984; Terborgh et al., 1990). The latter lists, both of ca. 540 spp., are based on inventories in areas of roughly 5,000 and 1,000 ha, respectively. The area surveyed at Alto Madidi over a period of only 14 days, consisted mainly of a transect ca. 6 km-long by 200 m, through young river-edge forest (ca. 100 m), mature floodplain forest (ca. 400 m), and older forest on somewhat hilly alluvial terraces (ca. 5.5 km).

In this small area we recorded 403 species of birds (Appendix 1). This number probably represents about 95 percent of the resident bird community. Based on more prolonged fieldwork at Tambopata and Cocha Cashu, we predict that an additional 75 species will eventually be found at Alto Madidi, including a

TABLE 1. VEGETATION TRANSECTS - LOWLAND MOIST FORESTS.
At each site, sum of 10 transects, each 2x50 m. Includes plants with stems diameter ≥ 2.5 cm. at breast height.

	# of fam.	Total spp.	Total ind.	Liana spp.	Liana ind.	Tree spp.	Tree ind.	Trees ≥ 10 cm dbh spp.	Tree ≥ 10cm dbh ind.
Avg. Moist Forest	46	152	373	35	68	116	304	42	64
Madidi	61	204	434	53	93	151	341	56	86
Madidi Ridge	49	175	483	44	85	131	398	64	89

small number of uncommon or rare residents, and a larger number of austral and Nearctic migrants. The absence of oxbow lakes and their marshes accounts for the few waterbirds on the list; the Tambopata and Cocha Cashu study areas include large oxbows with numerous resident and transient water and marshbirds.

The high avian diversity at Alto Madidi was not unexpected. It reflects both the habitat heterogeneity of the region, especially the combination of riverine and hill forest habitats in a small geographic area, as well as the general species richness of upper Amazonian forests near the base of the Andes (see Terborgh 1985 for a discussion of additional causes). A breakdown of Alto Madidi bird species diversity by habitat reveals that upland (*terra firme*) forest supports the richest community (182 spp.), followed by floodplain/river-edge forest (143 spp.), marsh/water birds (33 spp.), low second-growth (21 spp.), and migrants (20+ spp.), most of which occurred in river-edge forests. Four aerial species could not be assigned to a habitat. A comparison of bird diversity within *terra firme* and riverine forests in other parts of Amazonia (Table 2) shows the Alto Madidi forests to be equally diverse, and the total list of resident forest species to be almost as high as those available for the richest known sites.

The total bird list would be much higher if we had included lower montane forests (at 900-1200 m) on the ridges within 15 km to the west of the study site. At these elevations on the Serranía Pilón, a southerly extension of ridge just a few kilometers to the south of the Alto Madidi camp, Parker recorded (during fieldwork in June 1989) 43 bird species not listed in Appendix 1. Many additional species occur at even higher elevations to the west (see Appendix 4), and in natural grasslands along the Ríos Heath and Madidi to the east (see Appendices 2 and 3).

TABLE 2. BIRD SPECIES RICHNESS AT 16 AMAZONIAN FOREST LOCALITIES OF COMPARABLE SIZE. SPECIES TOTALS

Sites	Upland forest	Floodplain forest*	Total	Latitude	Reference
Limoncocha, EC	190+		513	2° 55'S	Pearson et al. 1977
Sucusari, PE	207		501	3° 16'S	Parker unpubl. list
Yanamono, PE	220		510	3°23'S	Parker unpubl. list
Vainilla, PE	201		328	3°46'S	Robbins et al. in press, Parker unpubl. data
Río Shesha, PE	216		360	8° 09'S	O'Neill et al. unpubl. list
Mucden, BO	207		288	11° 00'S	Remsen and Parker unpubl. list
Cocha Cashu, PE	-	217	526[1]	11° 51'S	Terborgh et al. 1984
Tambopata, PE	196	142	554	12° 36'S	Parker unpubl. list
Alto Madidi, BO	182	143	403	13° 10'S	Parker unpubl. list
Huanchaca, BO	192		-	14° 00'S	Bates and Parker unpubl. list
Cachoeira Nazaré, BR	236	82	447	10° 20'S	Stotz and Schulenberg unpubl. list
Fazendas Esteio P. Alegre, Diamona, BR	231			1° 59'S	Stotz and Bierregaard 1989
Reserva Ducke, BR	209		351	2°55'S	Willis 1977, Stotz in litt.
Raleigh Falls, SU	225		362	4° 50'N	Davis 1982, mimeographed list
Itaituba, Río Tapajos, BR	190+		350+	4° 50'S	Parker and Schulenberg unpubl. report
Altamira, Río Xingu, BR	150+		260	3° 20'S	Graves and Zusi 1990

*Additional species restricted to floodplain forest

1Upland species poorly known

Country abbreviations: BO-Bolivia, BR-Brazil, EC-Ecuador, PE-Peru, SU-Suriname.

We predict that more than 1,000 bird species, or an amazing 11 percent of all bird species on earth, will eventually be recorded along a transect from the Andean grasslands near Lake Titicaca to the lowland forests and savannas near the mouth of the Río Heath.

Our fieldwork at Alto Madidi reveals how little we know of bird distribution in northern Bolivia (and birds are the best-known vertebrate group!). Of 403 species found, nine were new to the Bolivia list, 30 were recorded in the country for only the second time (the first records of these were reported by Parker and Remsen in 1987), and 52 species were new for the department of La Paz.

Species new to Bolivia include an inconspicuous parrot (*Nannopsittaca dachilleae*) recently discovered in southern Peru (O'Neill et al., 1991), the little-known and uncommon Scarlet-hooded Barbet (*Eubucco tucinkae*), Crested Foliage-gleaner (*Automolus dorsalis*), Rufous-tailed Xenops (*Xenops milleri*), Undulated Antshrike (*Frederickena unduligera*), Ash-breasted Gnateater (*Conopophaga peruviana*), Olive-striped Flycatcher (*Mionectes olivaceus*), Black-and-white Tanager (*Conothraupis speculigera*), and Casqued Oropendola (*Clypicterus oseryi*). By far the most unexpected species observed during our days in lowland La Paz was an Arctic Tern (*Sterna paradisaea*) found and photographed at Laguna Santa Rosa along the lower Río Tuichi during our reconnaissance of that area on 24 May. Not only was this species previously unknown in Bolivia, it is also the first record of this pelagic tern for the interior of South America.

Interesting features of the resident avifauna at Alto Madidi included the presence of nine species of forest tinamous, an unusually large number for such a small area, and 16 species of parrots. Orange-cheeked Parrots (*Pionopsitta barrabandi*), although here reported for only the second time in Bolivia, were unusually common; we repeatedly observed small groups in the canopy and flying overhead, and occasionally noted as many as 12 in one flock. The abundance of this and several other parrot species may have been related to the presence of large numbers of fruiting fig trees (*Ficus sphenophylla* described above). Red-and-green Macaws (*Ara chloroptera*) were also common. Large cracids, including heavily hunted species such as Razor-billed Curassow (*Mitu tuberosa*) and Spix's Guan (*Penelope jacquacu*), were also relatively numerous.

The large number of antbirds (42 species) found in the study area reflects the mixing of foothill (e.g., *Myrmeciza fortis*) and riverine forest species (*Myrmeciza goeldii*). Some species, such as the Rufous-capped Antthrush (*Formicarius colma*), were inexplicably scarce, whereas others known from nearby areas to the north (e.g., *Myrmotherula ornata*) were not found at all.

The Alto Madidi avifauna, especially when combined with that of the montane forests close by to the west and that of pristine savannas not far to the northeast along the Río Heath (see Appendix 2), is unquestionably the richest known from any region in Bolivia, if not all of South America. Furthermore, about 10 percent of the bird species found in these areas are endemic to a relatively small (<100,000 km^2) section of southwestern Amazonia. This further underscores the conservation importance of the region.

Our fieldwork at Alto Madidi reveals how little we know of bird distribution in northern Bolivia . . . Of 403 species found, 9 were new to the Bolivia list, 30 were recorded in the country for only the second time . . . 52 species were new for the department of La Paz .

Mammals of Alto Madidi (L. Emmons)

In 12 days at this site, 45 species of mammals were identified (Appendix 4), which should represent about 50 percent of the non-flying mammal fauna and about 15-20 percent of the bats. This is a good result for the time spent and provides an accurate picture of the nature of the fauna.

The mammal fauna of the Alto Madidi region is similar to that of the Tambopata Reserve (82 species known; Emmons and Barkley, unpubl.), about 100 km to the northeast in adjacent Madre de Dios, Peru. Most of the differences between the two lists are likely to disappear when both faunas are more completely inventoried. However, there are two noteworthy species present at Tambopata but not found at Alto Madidi. Green Acouchys (*Myoprocta pratti)* would certainly have been seen by us if present, and were unknown to residents of the logging camp; and two-toed sloths (*Choloepus* sp.) also were unknown to the workers at Alto Madidi, although three-toed sloths were said to occur in the area. The Río Heath may be the southern distributional limit for these two species.

According to informants, White-lipped Peccaries (*Tayassu pecari*) are also absent from the area, as they may now be from the Tambopata Reserve, but they are said to occur far down the Río Madidi. Local opinion is that they used to occupy the whole region, but that they were exterminated by the *petroleros* (oil-drilling crews). However, in recent years white-lips have had an abrupt and severe population crash on the Río Manu, with epidemic disease as the most likely cause. As the status of the white-lipped peccary is unclear and locally endangered over much of Amazonia, it is important to report its presence or absence from given areas.

At the Alto Madidi site we found two species of mammals (apparently) previously unrecorded in Bolivia: a spiny tree rat (*Mesomys hispidus*), and a little big-eared bat (*Micronycteris nicefori*). That two species were added to the Bolivian list in only a few days suggests that a complete inventory of the area would turn up many more.

The community structure of mammals at Alto Madidi has some unusual features. The most striking is its extraordinarily high number of tapirs (*Tapirus terrestris*) and spider monkeys (*Ateles paniscus*), which respectively dominate the terrestrial and arboreal biomass. The four core RAP members have travelled extensively throughout the Neotropics, and none of us has previously seen an area with so much evidence of tapirs or a primate fauna so dominated by spider monkeys. Members of our group saw three tapirs and heard others, but their high numbers were largely inferred from incredible numbers of tracks throughout all parts of the forest, and the appearance of many tracks daily wherever we worked. The terrain at Alto Madidi is characterized by steep, eroded ridges with unstable soils that constantly slump downhill in mudslides and subsequently grow up in secondary scrub. It is possible that this mosaic of vegetation stages (clearly seen throughout the region during overflights) is highly favorable habitat for browsers such as tapirs. Another favorable circumstance is the absence of human inhabitants in the region, especially indigenous hunters, who tend to severely decimate tapir populations. The abundance of large game species at Alto Madidi suggests that the area has been little hunted for decades.

The primate fauna at Alto Madidi includes seven species, almost exactly the same as at the Tambopata Reserve, but depauperate when compared with the 13 species at Cocha Cashu Biological Station, Parque Nacional Manu, about 350 km to the north. We believe that the primates of the immediate area of the camp were all identified, because local hunters had not seen additional species. Five of the species were very abundant; only Squirrel Monkeys (*Saimiri sciureus*) and Red Howler Monkeys (*Alouatta seniculus*) were relatively uncommon. *Cebus albifrons* can be expected to occur in the region although it was not found near the camp. Expedition members made unconfirmed sightings of this species in forest on the ridge west of Ixiamas. Because spider monkeys have disappeared or are scarce in heavily hunted areas throughout Amazonia, the high numbers at Alto Madidi are of considerable importance for conservation.

There are large populations of Kinkajous (*Potos flavus*) and Olingos (*Bassaricyon gabbii)* at Alto Madidi. Both these and spider monkeys may benefit from the unusually high numbers of enormous, free-standing figs (*Ficus sphenophylla*) in this forest.

As is typical of most of lowland Amazonia, the terrestrial, nocturnal fauna is dominated by spiny rats (*Proechimys* spp.). Two species were collected, but the habitats were not adequately sampled for rodents, and other species of *Proechimys* may occur. Many other small mammals are to be expected.

Of special interest for conservation were two sightings, by Parker and Castillo, of a Short-eared Dog (*Atelocynus microtis*). This is one of the rarest mammals in Amazonia, with few reliable recorded sightings by scientists. In one sighting, the dog had a large frog in its mouth, which is the first record of food habits for a wild individual. Although short-eared dogs could occur in all lowland Amazonian parks or reserves from Colombia to Peru, and south of the Amazon in Brazil, there is apparently no information to confirm that they are actually present in any of them. It would therefore be of importance for the conservation of this species to preserve an area where they are definitely known to be present.

The forest 13 km west of Ixiamas seems to be an extension of lowland evergreen forest similar to that at Alto Madidi. The eight mammals recorded at this site are all common and widespread throughout western Amazonia. The area seen had clearly been under considerable subsistence hunting pressure for some time.

PAMPAS REGION

Parker and Castillo found 135 bird species in the savanna habitats at Ixiamas, including 40 species (29 percent) not previously recorded from the department of La Paz (see Appendix 3). All of these were known in Bolivia from only a few localities to the east and south in the department of Beni. These discoveries reflect the paucity of distributional data available for Bolivian grassland vertebrates. Of more than three large areas of grassland in lowland La Paz, not one had been previously inventoried; in fact, only one reliable bird list is available for the ca. 100,000+ sq km of natural, lowland grasslands in the country (see Remsen 1986). The ecological distributions of grassland bird species within the complicated mosaic of savanna plant communities (see Haase and Beck 1989) are largely unknown.

At Ixiamas we found several grassland bird species that are declining throughout most of their ranges in central South America (e.g., Cock-tailed Tyrant, *Alectrurus tricolor*, and Black-masked Finch, *Coryphaspiza melanotis*). That large populations of such species survive in northern and central Bolivia underscores the importance of establishing reserves of one kind or another in Bolivia.

Residents of Ixiamas reported that large areas of fairly pristine grassland and gallery forest, with few or no cattle, are situated in unpopulated areas between Ixiamas and the Río Beni to the east. Even large mammals (e.g., marsh deer) are said to survive in numbers there. Identifying and preserving (at least portions) of such areas should be given high priority.

One pristine grassland area of significant conservation importance is situated along the Bolivia/Peru border on the lower Río Heath. Our overflights of the Pampas del Heath revealed large expanses, possibly of as much as 10,000 hectares, of pantanal-like grasslands, and cerrado-like woodlands surrounded by very extensive riverine and upland evergreen forests. Vertebrate and plant surveys in these habitats on the Peruvian side of the river revealed high bird diversity (including the only

Peruvian populations of at least 14 grassland species), large populations of large mammals, and undisturbed grasslands and cerrado-like vegetation which differ floristically from those to the south (see Appendix 2; Parker, unpubl. data; Gentry, unpubl. data).

Gentry saw only one species of plant on the Ixiamas pampas that was held in common with the Peruvian pampas. We noted 57 families, 120 genera, and 169 species in the area. (See Appendix 10.) Beck observes that the Ixiamas pampa has little in common floristically with the savannas he has studied farther to the east. The initial indication is that these various savannas constitute a far more complicated mosaic of isolated habitat islands than a glance at the map would suggest.

The initial indication is that these various savannas constitute a far more complicated mosaic of isolated habitat islands than a glance at the map would suggest.

The Pampas del Heath is as yet almost unexplored for mammals, but as one of the only pristine grassland systems not yet degraded by cattle or severely hunted and too-frequently burned by man, it represents one of the chief conservation priorities and opportunities in Amazonian South America. Marsh Deer (*Blastocerus dichotomus*), extremely endangered or threatened over much of their geographic range, are found in the Pampas del Heath. There are also persistent rumors of Maned Wolves (*Chrysocyon brachyurus*). On our overflight we saw that the Bolivian pampas are criss-crossed with innumerable, deeply worn tracks of large mammals, presumably made by tapirs or deer. Parts of the savannas are densely studded with termite mounds, which should provide favorable habitat for anteaters and armadillos. Although its mammal list may not be large, the Pampas del Heath may be one of the only undisturbed natural habitats of its type and a refuge for species that are persecuted elsewhere and in need of protection.

APOLO REGION, MID ELEVATION WET FOREST (CALABATEA)

Physiography of Mid-elevation Yungas: Apolo to Calabatea (R. Foster)

The middle yungas is an extensive area of high relief mostly between 1,000 and 2,000 m, and Apolo is in the center of this region. The area west to the main Andes is geologically mapped as completely of Ordovician age. It is dissected with an irregular reticulate drainage from a labyrinth of twisting and turning ridges. To the east by contrast, the ridges are long, straight, and sharply defined, separated mostly by long gradual slopes and occasional vertical-walled mesas. These are mapped as narrow parallel strata of Ordovician, Devonian, Carboniferous, Cretaceous, and Tertiary age (not occurring in any predictable order). Apolo is in a broad, gently sloping valley bordered by a steep escarpment on its east flank and rolling hills to the west.

The significance of this contrast between the western and eastern parts of the northern yungas is not yet clear since we were only able to fly over the eastern part. From the air, however, the eastern part seemed to have considerably more diversity of vegetation types, presumably from the exposure of very different geological strata.

The Ordovician substrate, as seen in the road cuts in the Yuyu drainage to the south of Apolo and in exposure throughout the denuded Apolo Valley, is a wild variety of thin, soft strata tilted at a steep angle or even vertical. Most of these multi-colored layers are like soft clay— you can stick your fingers all the way in—and most of the rest are at best very brittle shale. The hardest strata seem to be the narrow bands of sandstone. These are also most likely to be on ridgetops.

This substrate seems remarkably permeable to water. For all the heavy rain we witnessed at the Calabatea Road Camp, the streams down the slopes hardly increased. Erosion seemed almost entirely caused by landslides except for along the large stream at the bottom of the valley. The Apolo valley floor is similarly surprising in the lack of stream erosion in spite of the heavy rains, severe denudation, and near absence of soil. Presumably the water percolates through the "rock." It would be instructive to find out how far the water table drops during periods of drought, and the effect this has on the plants.

The center of the middle yungas is probably the driest part. Layers of moist air coming from the east off the Amazon plain over the lower Tuichi valley and from the southwest up the deep river valleys of the Mapiri drainage are intercepted by the series of high ridges both east and west of the Apolo area.

Plant Communities (R. Foster)

The elevation range observed (1,000 to 1,650 m) is where the montane flora (e.g., *Clethra*, *Laplacea*, *Podocarpus*, *Schefflera*) meets the lowland flora (e.g., *Simarouba*, *Sloanea*, *Symphonia*, *Tachigali*). The overlap is considerable, with even the palms *Dictyocaryum lamarckianum* and *Iriartea deltoidea* growing side by side in places. Even the montane genera *Hedyosmum* and *Brunellia* were found to be more common at the valley bottom than on the ridges above. Gentry considers the distinction between montane and lowland forest types here to be more abrupt than I have presented. His sample transect of 0.05 ha at 1,500-1,550 m was comprised almost entirely of lowland taxa. Although Gentry could not sample the montane forest at Calabatea, he states that "a sample from Incahuara, farther south in the department of La Paz in the same ridge system, is available and appears to represent the same vegetation type."

High Forest

The best-developed forest (up to 30 m tall) is at the bottom of the valleys, and comes up the slopes especially on old landslide debris. Similarly, species richness is greatest in this forest but clearly decreases up slope to the ridge.

While tree ferns are common and epiphytes certainly more abundant than in lowland forest, there was not such a superabundance of epiphytes nor heavy moss cover to suggest year-round high humidity except on the very topmost ridges.

The most abundant tree appeared to be *Hyeronima* sp. (Euphorbiaceae), but the forest is most easily characterized by the numerous tree species of Rubiaceae (*Cinchona* and close relatives), Melastomataceae (*Topobea*, *Miconia*, etc.) and Lauraceae.

Stunted Forest

Forest on ridges is clearly shorter (less than 20 m), as is expected. But on large parts of the ridges the forest is stunted with a canopy often no more than 5 to 10 m tall. The easiest explanation is that the stunting (without much difference in flora) has to do with the water-holding capacity of the soil. However, nothing was seen to substantiate or refute this idea. It is possible that areas deliberately burned in exceptionally dry years regenerate in a stunted form. Several areas of recent burning to provide forage for pack animals were found along an old trail down the ridge. These are now growing back to bracken fern (*Pteridium aquilinum*) except in the more settled areas where the burning is continued to keep an area open for cattle pasture.

Palm Forest

While palm species in general are few in this area (six seen), some ridges and slopes are dominated by the conspicuous emergent palm, *Dictyocaryum lamarckianum*. These areas are so conspicuous that from a good vantage point with binoculars one can easily pick out all the patches for a distance of many kilometers. The area of palm forest appeared to be somewhere between 5 and 10 percent in this region. Some isolated large patches of *Dictyocaryum* palm forest were seen from the air northeast of Apolo, half-way to the Río Tuichi. The areas where these palms dominate are all associated with a sandstone substrate. This corresponds to personal observations in the Yanachaga and Pantiacolla mountains of Peru, that stands of *Dictyocaryum* are associated with sandy, acid substrate.

Bamboo Forest

Slopes with high concentrations of bamboo (*Guadua*, different from the lowland species) are occasional and not extensive in the Yuyu drainage area. But in the overflight of the high valleys to the northeast near the Tuichi, one could see many areas of several square kilometers dominated by bamboo. It is not immediately obvious what sort of disturbance lead to the invasion of bamboo. It does not regenerate in the shade of closed canopy and only rarely establishes in single treefall gaps. The skinny, clambering bamboo, *Chusquea*, does, however, take advantage of these smaller gaps, and is common in the ridgetop forest.

Apolo Matorral

In the Apolo valley and the rolling hills and ridges paralleling it on either side, the only forest remains are in small pockets in the hills both in ravines and up on slopes. The rest is burned frequently, though there is much exposed substrate and the vegetation is a thin cover of grass and other herbs mixed with gnarled shrubs and small trees, notably an abundant *Alchornea* (Euphorbiaceae). Closer to Calabatea the matorral is more lush, with *Didymopanax morototoni* (Araliaceae) scattered throughout. However, this area is far from the population center and presumably has suffered less human intervention and for a shorter time.

Although almost certainly drier than the Calabatea forest, it is striking that the forest patches that remain in the Apolo matorral are almost identical in composition (though with fewer species) to that wet forest. According to Beck, the grasses are nearly all species known from the pampas at a much lower elevation. Among the shrubs and small trees are a few species also found in the patch of cerrado vegetation at Chaquimayo. There appears to be little in the flora that is unique or distinctive, which is surprising given the unusual and isolated vegetation type over this large an area. I would argue that the long history of human habitation in this area, pre-Columbian as well as the last few centuries, has created this vegetation from what was originally a moist forest not too different from that at Calabatea. Gentry argues on the contrary that it is "likely that much of the now ecologically devastated area around Apolo was originally cerrado to judge from a few isolated trees of *Tabebuia ochracea* and *Cybistax antisyphilitica* that we saw en route to Calabatea."

Phytogeography (A. Gentry)

The floristic composition of the Calabatea sample is distinctly unexpected in view of the southerly latitude of Bolivia. Since there is a general tendency for montane plant communities to be lower on mountains farther from the equator, one would expect lowland tropical elements to be restricted to altitudes near sea level. Instead, the opposite occurs and these lowland plants, both the individual species and the higher taxa, appear to occur at higher elevations than elsewhere in Latin America. Records of lowland families like Caryocaraceae, Quiinaceae, and Humiriaceae, genera such as *Roucheria*, *Curarea*, and *Callichlamys*, and species like *Iriartea deltoidea* above 1,500 m appear to represent new altitudinal extremes. The potential conservational significance of this surprising result, if any, remains to be elucidated, but its discovery, especially if borne out as a general pattern, is of more than trivial theoretical significance.

Prior data from a second, higher elevation site at Sacramento Alto (2,500 m) in the Coroico valley, department of La Paz, are also available and are of interest in comparison with data from Calabatea and Incahuara. At this altitude, the switch from lowland to montane taxa has been completed with a concomitant loss in species diversity to 91 spp. 2.5 cm dbh in 0.1 ha. Melastomataceae (15 spp.) and Rubiaceae (10 spp.) are the most speciose families, just as they are at Calabatea. With a reduced number of species, Lauraceae (7 spp.) share canopy dominance with such montane trees as species of *Alchornea*, *Clethra*, *Gordonia*, *Hedyosmum*, *Meliosma*, *Myrsine*, *Rhamnus*, *Symplocos,* and *Weinmannia.* At this altitude, Compositae completely dominate the liana component of the flora (7 of 16 free-climbing species). Forests such as this undoubtedly occur above 2,000 m on the slopes west of Calabatea.

Plant diversity (R. Foster and A. Gentry)

(Foster): The floristic richness observed here is nothing exceptional (we noted 113 families, 251 genera, and 390 species — see Appendix 8). It is certainly far less than forest at similar elevations in southern and central Peru. There was no clearly dominant large tree, but the same one or two dozen common tree species could usually be found throughout the area from ridgetop to ravine bottom, from high forest to stunted forest, and from palm forest to bamboo forest. Even the remnant pockets of forest on the hills adjacent to Apolo many kilometers away had almost the same composition. The unimpressive diversity and uniformity is surprising given that this is the transition zone between montane and lowland floras with consequent mixing of species from each.

(Gentry): The 110 spp. in 500 m^2 at Calabatea implies that at least 140-150 spp. occur in 0.1 ha, about the same as in the Incahuara sample (147 spp.). Because of the incomplete sample from Calabatea, only a very broad outline can be suggested with respect to species diversity and floristic composition. The two most diverse families are melastoms and rubiacs (ca. 12 spp. each), with legumes (9 spp.), Lauraceae and Moraceae (6 spp. each), and Bignoniaceae (4 spp.) following as the most speciose. The forest at Incahuara is characterized by a predominance of melastoms (18 spp.) and Rubiaceae (12 spp.) in the understory and lower canopy, but differs in the overwhelming diversity of Lauraceae in the canopy (24 spp. in 0.1 ha -- the greatest Lauraceae diversity yet found in my 0.1 ha samples throughout the world, and the majority of the species are likely undescribed). How the extreme prevalence of this preeminently bird-dispersed family might be related to the local avifauna or other elements of the biota remains an unexplored question.

Birds of Mid-elevation Wet Forest (Calabatea) (T. Parker)

The Calabatea avifauna is comprised of a mixture of highland and lowland species (Appendix 4). Of 169 species found between ca. 1,300 and 1,600 m, 80 species are montane, most occurring in low- and mid-elevation forests (900-1,500 m) from Colombia south to central Bolivia. The rest are lowland species which were at or near the upper limits of their elevational ranges. Ten of the montane species are endemic to the yungas region of extreme southern Peru/northern Bolivia, as are numerous well-marked subspecies. Most of the montane bird species found at Calabatea occurred primarily above 1,500 m in forests characterized by numerous highland plant genera (see above).

Bird species diversity in the Calabatea forest is similar to that found in the only other well-worked cloud forest site at the same elevations in Bolivia: Serranía Bellavista, department of La Paz, where 184 species were recorded during >60 days of fieldwork (Remsen, unpubl.). Comparable lists from Ecuadorian and Peruvian cloud forests contain slightly higher totals (Davis 1986, Parker and Parker 1982, Robbins et al. 1987, Schulenberg et al. 1984, Terborgh and Weske 1979, Fitzpatrick et al., unpubl.), but sampling biases are difficult to interpret. Based on available data it seems safe to say that lower and middle Andean forest bird communities are very similar in structure and species composition along the entire eastern slope from southeast Colombia to northwest Bolivia.

The two most speciose bird families represented at Calabatea are Tyrannidae (32 species) and Emberizidae (42 species). Tanagers (especially 9 *Tangara* spp.) comprised a high percentage of total individuals in canopy flocks. Two species not previously reported from Bolivia were found, the Sharpbill (*Oxyruncus cristatus*) and Chestnut-breasted Wren (*Cyphorhinus thoracicus*). We were unable to reach the tallest ravine forests at Calabatea, where additional (important) lower montane species might have been found (e.g., *Terenura sharpei*, *Odontorchilus branickii*, and *Tangara chrysotis*).

In summary, the Calabatea forests support a rich but fairly typical lower montane/upper tropical avifauna. About 10 percent of the taxa are endemic to the yungas region of northern Bolivia, and this number will be even larger in upper montane forests to the west. Humid montane forests, especially at lower elevations (800-1,500 m), are in urgent need of protection. Tall, montane forests at these elevations to the east and south (e.g., above the Ríos Beni and Tuichi) probably support somewhat richer, but very similar avifaunas.

APOLO REGION, MID ELEVATION DRY FOREST (CHAQUIMAYO, VALLE DEL MACHARIAPO)

Physiography of the Machariapo Valley (R. Foster)

The Machariapo valley lies to the northwest of Apolo and drains north into the Alto Tuichi. It is deep (1,000 m at the bottom) but fairly broad, with steep slopes, above which are high plateaus punctuated with rounded peaks. The rock underlying the valley is geologically mapped as of Devonian age. The exposed rock visible along the trails and river-cuts is radically different from that on the Ordovician of Apolo and the area southwest. Instead of a "baklava," pastel-colored assortment of different thin layers, the head of the valley (Chaquimayo) is uniformly covered with dark grey, hard slate. A few kilometers down the valley this changes into a light colored conglomerate of rounded rocks, not associated with the current floodplain but going way upslope. The latter is similar in appearance to the old alluvium bordering the river draining the center of La Paz. However, no difference in vegetation was seen between the two rock types.

Plant Communities (R. Foster and A. Gentry)

Deciduous Forest

Because the valley is mostly walled in on all sides, it is truly drier than the nearby Apolo valley. Nevertheless, given the overwhelming evidence of previous forest cutting, burning, and coffee planting in the understory, the completely deciduous character of the valley slopes must be seen as an artifact of human influence favoring the deciduous species. Left to itself, the forest would probably return to a more semi-deciduous character, with the slower-growing, deep-rooted species moving back in from their refuges in the ravines and river banks. The forest in these latter areas is nearly evergreen.

The deciduous forest continues down the valley as far as the eye can see. It is probable that the Alto Tuichi to which this valley drains has similar vegetation, as would the valleys parallel to and north of the Machariapo on into Peru. It is important to explore this possibilty on the chance there are some valleys with less-disturbed vegetation.

The overwhelmingly dominant tree (in the transect sample, half of all trees 20 cm dbh and two-thirds of all trees 30 cm) is *Anadenanthera colubrina*, a mimosoid legume with a distinctive spreading crown "noted for its hallucinogenic indole alkaloids." Also abundant in the canopy is an *Acacia*, *Astronium*, *Schinopsis*, and a short *Ceiba*. Understory trees include a conspicuous large *Echinopsis*? cactus, *Triplaris*, and *Capparis*. Distinctive and common shrubs include the startling combination of a strange *Opuntia* (60 plants in 0.1 ha, by far the most common species in the sample) with small, flattened joints and straight, erect stem occasionally reaching 12 cm dbh and 12 m or more tall, mixed with *Clavija* and a small *Trichilia*. Epiphytic orchids and bromeliads are abundant. The evergreen patches include typical lowland moist forest trees such as *Gallesia*, *Platymiscium*, *Myroxylon*, *Clarisia biflora*, *Cecropia polystachya*, *Ficus juruensis*, etc.

Structurally this forest is not very different from moister forests, with 465 stems 2.5 cm dbh including 77 trees, 2 lianas, and a strangler 10 cm dbh. The most striking structural anomaly is the presence of 134 individual lianas (compared to an average of 71 for all dry forests sampled) probably relating to a history of past disturbance.

Scattered individuals of commercially important timber taxa such as *Amburana cearensis*, *Tabebuia impetiginosa*, and *Cedrela* suggest that the forest may formerly have been more diverse. *Myroxylon balsamum*, usually fairly rare, is common here, and forms the basis for a local "balsam of Peru" industry (we shared our charter flight from Apolo to La Paz with a cargo of *Myroxylon* bark). Since the species is so common, it might form the basis of a forest product economy.

Ridge Savanna-Cerrado

On the promontory ridge at the head of the valley (1,500 m), the deciduous forest on the steep slope changes rather abruptly into an unusual savanna-like vegetation (isolated trees or small clusters of trees surrounded by a matrix of grassland) as the slope starts to flatten out. Above this narrow area, on the divide over to Apolo, the terrain turns back into the denuded soil with isolated pockets of stunted wet forest vegetation typical of the rest of the Apolo valley.

This small anomolous area of only a few hectares seems to have a combination of small tree species known from the periodically flooded savannas of the Beni in Bolivia and the dry, open-forest cerrado vegetation further east in Brazil. The most abundant tree is a dwarf *Terminalia* sp., mixed in the open, regularly burned areas with *Jacaranda*, *Pseudobombax*, *Tabebuia*, *Byrsonima*, and in the thickets with *Diospyros* and *Dilodendron*. There is almost no floristic overlap with the adjacent dry forest.

Binocular observation of the surrounding rim of the valley did not reveal any more areas like this one, though there were possibly a few very small patches.

Other

On a sloping mesa high up one side of the valley, an estimated 90 percent or more of the trees are an evergreen *Vochysia*—presumably the result of former human clearing on a substrate different from that of the valley slopes.

Diversity and Phytogeography (A. Gentry and R. Foster)

A 0.1 ha sample of this forest confirmed that it has many typical dry forest features, such as relatively low diversity (we noted 71 families, 190 genera, and 275 species.—see Appendix 9), extreme prevalence of legume trees and bignon vines, and the occurrence of families such as Cactaceae and Capparidaceae. The 79 species 2.5cm dbh in 0.1 ha is quite typical of lowland Neotropical dry forests, which average 60 spp. in similar samples. The 29 families represented is comparable to the average of 25 represented in similar samples. (See Table 3.)

This isolated dry forest includes a number of taxa new to Bolivia that may have conservation importance, including two lianas known previously from *eastern* Paraguay and southern Brazil—*Mansoa difficilis* and *Arrabidaea selloi*. Near the river, the American staghorn fern, *Platycerium andinum*, was not uncommon on trunks. This is similar in habitat to its previously known locations in semi-deciduous forest in central Peru where it was supposed to be restricted. It is the only Neotropical species of staghorn fern. It is likely that some plants such as the two dominant cacti (still unidentified) are endemic to Bolivia, if not to these very dry forest valleys.

Even though the species in the small patch of cerrado savanna on the ridge are widespread outside Bolivia, they are mostly poorly known within the country and all but one of the species for which there are distributional data are new to La Paz department. *Dilodendron bipinnatum*, for example, was previously known from four Bolivian collections, none from La Paz department; our collection of female flowers is the first for the species. The distribution and abundance in the yungas of this unexpected vegetation type is currently unknown, and presumably rare (though Beck notes that there are interesting savanna elements in degraded vegetation around Caranavi and Coroico).

Birds of Mid-elevation Dry Forest (Machariapo Valley) (T. Parker)

The montane, deciduous forests along the Río Machariapo north of Apolo at ca. 1,000 m contain an interesting mixture of species, especially cloud forest forms not previously known from such a dry area, and a smaller number of the more expected dry forest species. The most unusual bird species found (as identified on tape-recordings made by L. Emmons and E. Wolf), appears to be an undescribed species of *Herpsilochmus* antwren, probably an isolate of the *atricapillus* group. The discovery of this species, along with our sightings of the Green-capped Tanager (*Tangara meyerdeschauenseei*), until now known from only one valley in Puno, Peru, in dry scrub near Apolo, suggests that other endemic bird taxa with very restricted ranges probably occur in the dry, upper portions of the Río Tuichi and similar river drainages in northern La Paz. Additional inventories of the flora and fauna of these valleys are urgently needed.

TABLE 3. VEGETATION TRANSECTS – MID-ELEVATION DRY FORESTS

	# of fam.	Total spp.	Total ind.	Liana spp.	Liana ind.	Tree spp.	Tree ind.	Trees ≥ 10cm dbh spp.	Trees ≥ 10cm dbh ind.
Dry For. Avg.	25	60	294	13	71	47	221	24	51
Chaquimayo	29	79	465	29	134	50	431	29	7

Mammals of Mid-elevation Dry Forest (Machariapo valley) (L. Emmons)

Five of the six mammals identified in the dry forest at Río Machariapo are species known from lowland evergreen rain forest. However, the overwhelmingly dominant rodent in the area, *Oryzomys nitidus*, (12 captures), is usually rare in rain forest. Its optimal habitat may therefore be dry forest, a fact that may not have been previously recorded. It is probably the chief wild vector of plague (*Yersinia pestis*), which is endemic around Apolo. Another rodent, *Akodon aerosus*, was found both in the wet cloud forest at Calabatea and in the dry forest along the Río Machariapo. It is typical of the eastern slopes of the Andes at around 1,000 m.

The most unusual mammal found at Machariapo was a bushy-tailed opossum (*Glironia venusta*), which was seen emerging from a tree hole and climbing to the canopy in a patch of older dry forest with a well-developed understory. This forest appeared to have been free from fire for longer than most of the other dry forest we saw. *Glironia* is extremely rare and known from only eight individuals. One specimen was collected in "Yungas, Bolivia" in 1867. All individuals from known habitats are from lowland evergreen rainforest, so it is noteworthy to have seen it in dry forest at 1,000 m.

The Río Machariapo area was under intense hunting pressure. Hunters with dogs scoured the valley bottom on two of the three days we spent there and we heard several gunshots. Nonetheless, the dense spiny undergrowth provides some protection for large terrestrial mammals, and deer and agoutis persist along the river.

CONCLUSION

This one-month assessment of the biological resources of northwest Bolivia represents a starting point for the conservation of the region. Due to its global importance, the area clearly merits international support for exploration, research, and conservation. A concerted, prompt effort to study and protect Madidi and environs could make a lasting contribution to the world's natural heritage bank. In an era of unprecedented land alteration worldwide, such opportunities are rare.

Appendices

Codes for Avian Data

Habitats

A	Aguajales; groves of palms (*Mauritia flexusosa*) occurring in poorly drained grassland areas, or along the edges of oxbow lakes.
Gf	Gallery forest; bordering streams or occurring as isolated patches on higher ground in otherwise open grassland. Floristically distinct from taller, continuous forest on terra firme.
Lm	Lake margin; the low aquatic vegetation growing in mats as well as the narrow fringe of shubbery and small trees that border oxbow lakes.
P	Pantanal-like grasslands; These seasonally flooded savannas are often dominated by a diversity of sedges, with lesser numbers of gras species; clumps of bushes and small trees occur throughout all but the wettest areas.
Fh	Mature forest on well-drained, high ground (*terra firme*)
Ft	Floodplain or "transitional forest"; tall forest that is seasonally inundated in places by overflow from river and/or rainfall
Fs	Swamp forest; permanently flooded forest within transitional forest
Fe	Forest edges
Fsm	Forest stream margins
Fo	Forest openings (primarily treefall gaps)
Z	"Zabolo"; riverbank forest characterized by trees such as *Cecropia*, *Ochroma*, and *Erythrina*, with an undergrowth of *Gynerium* cane, *Guadua* bamboo, and broad-leafed plants including *Costus* and *Heliconia*.
B	Bamboo (*Guadua*) thickets within transitional or riverbank forest
C	Clearing; the large grassy clearing with scattered trees and bushes around the Aserradero Moira buildings and along the airstrip at Alto Madidi; bordered by a fringe of secondary woodland and extensive, tall transitional forest
R	River; the open water of the Río Madidi
Rm	River margins; vegetation overhanging riverbank, and fallen trees and rubble washed ashore during floods
S	Shores, sandbars, and rock outcrops along the river
M	Marsh; permanently flooded areas filled with grasses (*Paspalum*) and other water-adapted plants; in the clearing and at the end of the airstrip at Alto Madidi
O	Overhead (for aerial foragers); letters in parentheses following this code refer to habitats in which the species is most apt to occur

Foraging Position

T	Terrestrial
U	Undergrowth or understory (up to 5 m in tall forest)
Sc	Subcanopy or middlestory (mainly from 5 to 15 m in tall forest)
C	Canopy (primarily above 15 m in tall forest)
W	Water
A	Aerial

Sociality

S	Solitary or in pairs
G	Gregarious; large groups of same species (more than 5 individuals)
M	Mixed-species flocks
A	Army ant followers

Abundance

C	Common; recorded daily in preferred habitat in moderate to large numbers (e.g., more than 10 individuals along ca. 5.0 km surveyed daily)
F	Fairly common; recorded daily in small numbers (e.g., fewer than 10 individuals)
U	Uncommon; recorded every other day in small numbers
R	Rare; recorded fewer than 5 times
(M)	Migrant, origin unknown
(Mn)	Migrant from the north, primarily from North America, normally occurring only from mid-August to March
(Ms)	Migrant from south (April to October)

Evidence

sp	Specimen obtained in survey area
t	Tape-recording obtained in area
ph	Photo
si	Species identified by sight or sound
*	First record for La Paz departmant
+	First record for Bolivia

APPENDIX 1

Birds of the Alto Madidi Area

Theodore A. Parker, 1990

	Habitats	Foraging	Sociality	Abundance	Evidence
TINAMIDAE (9)					
Tinamus major	Ft	T	S	F	t
T. guttatus	Fh,Ft	T	S	F	t*
T. tao	Fh	T	S	F	t
Crypturellus cinereus	Ft	T	S	C	t
C. soui	Ft	T	S	F	t
C. obsoletus	Fh	T	S	U?	si
C. undulatus	Z,Ft	T	S	C	t
C. variegatus	Fh	T	S	R	t*
C. bartletti	Ft	T	S	U	t
PHALACROCORACIDAE (1)					
Phalacrocorax brasiliensis	R	W	S,G	U	si
ANHINGIDAE (1)					
Anhinga anhinga	R	W	S	R	si
ARDEIDAE (8)					
Ardea cocoi	S	W	S	U	si
Egretta alba	S	W	S	U	si
E. thula	S	W	S	U	si
Butorides striatus	M	W	S	R	si
Ardeola ibis	C,S	T	S,G	F	si
Pilherodias pileatus	S,M	W	S	F	si
Nycticorax nycticorax	S	W	S	R	si
Tigrisoma lineatum	M,Fsm	W	S	U	t
CICONIIDAE (2)					
Mycteria americana	S	W	G,S	U	si
Jabiru mycteria	S	W	S	R	si
THRESKIORNITHIDAE (1)					
Mesembrinibis cayennensis	Fsm	T,W	S	U	si
ANHIMIDAE (1)					
Anhima cornuta	M,S	T	S	F	t
Chauna torquata	M	T	S	R	si*
ANATIDAE (2)					
Neochen jubata	S	T	S	U	si
Cairina moschata	R,M	W	S	U	si
CATHARTIDAE (3)					
Sarcoramphus papa	Ft	T	S	F	si
Coragyps atratus	S,Z,Ft	T	G	U	si
Cathartes melambrotus	Z,Ft,Fh	T	S	F	si
ACCIPITRIDAE (13)					

APPENDIX 1

	Habitats	Foraging	Sociality	Abundance	Evidence
Gampsonyx swainsonii	Rm,C	A	S	R	si*
Elanoides forficatus	Ft,Fh	A	S,G	R(M?)	si
Leptodon cayanensis	Ft,Fh	C,Sc	S	U	si
Harpagus bidentatus	Fh	Sc	S	U	t
Accipiter bicolor	Ft	U,Sc	S	R	si
Buteo magnirostris	Rm,Z,C	T,U,Sc	S	F	t
Asturina nitida	Rm,Z,C	?	S	R	si*
Leucopternis albicollis	Fh	C	S	U	si
L. schistacea	Fs,Ft	T,U	S	U	t
Buteogallus urubitinga	Rm,S	T,U	S	U	si
Morphnus guianensis	Ft,Fh	Sc,C	S	R	si*
Spizaetus tyrannus	Z,Ft	Sc,C	S	U	t
Geranospiza caerulescens	Ft	T,U,Sc	S	U	si
PANDIONIDAE (1)					
Pandion haliaetus	R	W	S	R(Mn)	si
FALCONIDAE (6)					
Herpetotheres cachinnans	Ft,Rm,C	T,C	S	F	t*
Micrastur ruficollis	Fh	U,Sc	S	R	si
M. gilvicollis	Ft,Fh	U,Sc	S	U	t
Daptrius ater	Rm,S	T,C	S,G	U	t
D. americanus	Ft,Fh	Sc,C	G	F	t
Falco rufigularis	Rm,C	A	S	F	t
CRACIDAE (4)					
Ortalis motmot	Z,Ft	Sc,C	G	C	t
Penelope jacquacu	Ft,Fh	C,Sc,T	S,G	F	t
Aburria pipile	Ft,Z	C	S	U	si
Mitu tuberosa	Ft,Fh	T	S	R	t
PHASIANIDAE (1)					
Odontophorus stellatus	Ft,Fh	T	G	U	t
OPISTHOCOMIDAE (1)					
Opisthocomus hoazin	Rm	U,Sc,C	G	C	si
PSOPHIIDAE (1)					
Psophia leucoptera	Fh	T	G	U	si
RALLIDAE (5)					
Rallus nigricans	M	T,W	S	R	si*
Aramides cajanea	Ft,Fsm	T	S	F	si
Laterallus exilis	M,C	T,U	S	F	t
L. melanophaius	M	T,W	S	F	si*
Porphyrula martinica	M	T,W	S	U	si

Habitats	
Fh	Upland forest
Ft	Floodplain forest
Fs	Swamp forest
Fsm	Forest stream margins
Fo	Forest openings
Z	"Zabolo"
B	Bamboo
C	Clearing
R	River
Rm	River margins
S	Shores
M	Marsh
O	Overhead
Foraging Position	
T	Terrestrial
U	Undergrowth
Sc	Subcanopy
C	Canopy
W	Water
A	Aerial
Sociality	
S	Solitary or in pairs
G	Gregarious
M	Mixed-species flocks
A	Army ant followers
Abundance	
C	Common
F	Fairly common
U	Uncommon
R	Rare
(M)	Migrant
(Mn)	Migrant from north
(Ms)	Migrant from south
Evidence	
sp	Specimen
t	Tape
ph	Photo
si	ID by sight or sound
*	First record for La Paz
+	First record for Bolivia

APPENDIX 1

	Habitats	Foraging	Sociality	Abundance	Evidence
EURYPYGIDAE (1)					
Eurypyga helias	Ft,Fsm	T,W	S	U	si
CHARADRIIDAE (2)					
Hoploxpterus cayanus	S	T	S	F	si
Charadrius collaris	S	T	S	U	t
LARIDAE (2)					
Phaetusa simplex	R	W,A	S	U	si
Sterna superciliaris	R	W,A	S	F	si
RHYNCHOPIDAE (1)					
Rynchops nigra	R	W	S	R	si
COLUMBIDAE (7)					
Columba cayennensis	Rm,Z,C	U,Sc,C	S,G	U	si
C. subvinacea	Ft,Fh	C	S	C	t
Columbina talpacoti	C	T	S,G	F	t
C. picui	C,Rm	T	S,G	U(Ms?)	si
Claravis pretiosa	Ft	T	S	R	si
Leptotila rufaxilla	Z,C	T	S	F	t
Geotrygon montana	Ft,Fh	T	S	F	t
PSITTACIDAE (17)					
Ara ararauna	Ft	C	G	U	t
A. macao	Ft,Fh	C	S,G	F	t
A. chloroptera	Ft,Fh	C	S,G	F	t
A. severa	Ft,Z	C	S,G	F	t
A. manilata	Ft	C	G	R	si
Aratinga leucophthalmus	Z, Ft	C	G	U	t
A. weddellii	Z,Ft,C	C	G	C	t
Pyrrhura picta	Fh	C	G	F	t
P. rupicola	Ft,Fh	C	G	U	t
Forpus sclateri	Ft,Fh	C	S	U	t
Brotogeris cyanoptera	Ft,Fh,Z	C	G	C	t
Nannopsittaca dachilleae	Ft	C	S	R	t+
Pionites leucogaster	Ft,Fh	C	G	C	t
Pionopsitta barrabandi	Fh	Sc,C	S,G	F	t
Pionus menstruus	Ft,Z	C	S,G	U	t
Amazona ochrocephala	Ft,Z	C	S	U	t
A. farinosa	Ft,Fh	C	S,G	F	t
CUCULIDAE (8)					
Coccyzus melacoryphus	C,Z	C	S,M	U(Ms)	si
Piaya cayana	Z,Ft,Fh	Sc,C	S,M	F	t
P. melanogaster	Fh,Ft	C	S,M	U	t*

APPENDIX 1

	Habitats	Foraging	Sociality	Abundance	Evidence
P. minuta	C	U	S	R	t
Crotophaga ani	C	U,T	G	F	t
Tapera naevia	C	U	S	R	si
Dromococcyx phasianellus	Ft	T,U	S	R	t
D. pavoninus	Fh	U	S	R	t
STRIGIDAE (7)					
Otus choliba	Z,C	Sc,C	S	R	si
O. watsonii	Ft,Fh	Sc,C	S	C	t
Lophostrix cristata	Ft,Fh	Sc,C	S	U	t
Pulsatrix perspicillata	Ft	Sc,C	S	U	t
Glaucidium minutissimum	Ft,Fh	Sc	S	F	t
G. brasilianum	Z,C	Sc	S	U	t
Ciccaba virgata	Ft,Fh	C,Sc	S	U	si
NYCTIBIIDAE (1)					
Nyctibius grandis	Ft	C,A	S	U	t
CAPRIMULGIDAE (4)					
Chordeiles rupestris	S	A	G	C	si
Nyctidromus albicollis	Z,C	A	S	F	t
Nyctiphrynus ocellatus	Ft,Fh	A,U	S	U	t
Hydropsalis climacocerca	Rm,S	A	S	F	si
APODIDAE (5)					
Streptoprocne zonaris	O	A	G	F	si
Chaetura cinereiventris	O,Fh	A	G,M	C	t
C. egregia	O	A	G,M	U(Ms?)	t?
C. brachyura	O,Fh	A	G,M	R	si
Panyptila cayennensis	O,Fh	A	S,M	U	t*
TROCHILIDAE (13)					
Glaucis hirsuta	Z,Ft	U,Sc	S	U	t
Threnetes leucurus	Z,Ft	U	S	U	t
Phaethornis superciliosus	Fh	U	S	F	t
P. hispidus	Z,Ft	U	S	F	t
P. ruber	Ft,Fh	U	S	C	t
Florisuga mellivora	Ft,Fh	Sc,C	S	U	si*
Anthracothorax nigricollis	Rm,C	C	S	U	si
Thalurania furcata	Ft,Fh	U,Sc	S	F	t
Hylocharis cyanus	Ft	U,Sc,C	S	F	t
Amazilia lactea	C,Ft	Sc,C	S	R	si
Polyplancta aurescens	Ft,Fh	Sc,C	S	F	t*
Heliothryx aurita	Ft,Fh	Sc,C	S	U	si
Heliomaster longirostris	Z,C	C	S	R	si

Habitats	
Fh	Upland forest
Ft	Floodplain forest
Fs	Swamp forest
Fsm	Forest stream margins
Fo	Forest openings
Z	"Zabolo"
B	Bamboo
C	Clearing
R	River
Rm	River margins
S	Shores
M	Marsh
O	Overhead
Foraging Position	
T	Terrestrial
U	Undergrowth
Sc	Subcanopy
C	Canopy
W	Water
A	Aerial
Sociality	
S	Solitary or in pairs
G	Gregarious
M	Mixed-species flocks
A	Army ant followers
Abundance	
C	Common
F	Fairly common
U	Uncommon
R	Rare
(M)	Migrant
(Mn)	Migrant from north
(Ms)	Migrant from south
Evidence	
sp	Specimen
t	Tape
ph	Photo
si	ID by sight or sound
*	First record for La Paz
+	First record for Bolivia

APPENDIX 1

	Habitats	Foraging	Sociality	Abundance	Evidence
TROGONIDAE (6)					
Pharomachrus pavoninus	Fh	Sc,C	S	U	sp
Trogon melanurus	Ft,Fh	Sc,C	S	F	t
T. viridis	Ft,Fh	Sc,C	S,M	U	t
T. collaris	Ft,Fh	U,Sc	S,M	F	t
T. curucui	Ft,Z	Sc,C	S,M	U	t
T. violaceus	Ft, Fh	Sc,C	S,M	F	t
ALCEDINIDAE (5)					
Ceryle torquata	Rm	W	S	F	t
Chloroceryle amazona	Rm	W	S	F	t
C. americana	Fsm,Rm	W	S	U	si
C. inda	Fsm	W	S	U	si
C. aenea	Fsm	W	S	U	si
MOMOTIDAE (3)					
Electron platyrhynchum	Ft,Fh	Sc	S	C	t
Baryphthengus martii	Ft	Sc	S	F	t
Momotus momota	Ft,Fh	Sc	S	U	t
GALBULIDAE (2)					
Galbula cyanescens	Ft,Fo	U,Sc	S,M	F	t
Jacamerops aurea	Ft,Fh	Sc,C	S	U	t
BUCCONIDAE (7)					
Notharchus macrorhynchus	Ft,Fh	C	S	U	t*
Nystalus striolatus	Ft,Fh	Sc,C	S	U	t
Malacoptila semicincta	Ft,Fh	U	S	U	si
Nonnula ruficapilla	Ft,Z,B	U,Sc	S	U	si
Monasa nigrifrons	Z,Ft,C	Sc,C	G,M	C	t
M. morphoeus	Ft,Fh	Sc,C	G,M	C	t
Chelidoptera tenebrosa	Rm,C,Z	A	S	F	t
CAPITONIDAE (3)					
Capito niger	Ft,Fh	Sc,C	S,M	F	t
Eubucco richardsoni	Ft,Fh	Sc,C	S,M	F	t
E. *tucinkae*	Z,Ft	C,Sc	S,M	R	t+
RAMPHASTIDAE (8)					
Aulacorhynchus prasinus	Ft,Z	C	S	U	t
Pteroglossus castanotis	Ft,Z	C	G	F	t
P. inscriptus	Ft,Z	Sc,C	G	R	t*
P. mariae	Ft,Fh	C	G	F	t
P. beauharnaesii	Ft,Fh	C	G	U	t
Selenidera reinwardtii	Ft, Fh	Sc,C	S	F	t*
Ramphastos culminatus	Ft,Fh	C	S,G	C	t

APPENDIX 1

	Habitats	Foraging	Sociality	Abundance	Evidence
R. cuvieri	Ft,Fh	C	S,G	C	t
PICIDAE (15)					
Picumnus borbae	Ft,Fh	Sc,C	S,M	U	si
Chrysoptilus punctigula	Z,C	Sc,C	S	R	t
Piculus leucolaemus	Ft,Fh	C	M	F	sp
P. chrysochloros	Ft	C	S,M	U	t?
Celeus elegans	Ft	Sc,C	S,M	R	t
C. grammicus	Fh	C	S,M	F	t
C. flavus	Ft	Sc,C	S,G	U	t
C. spectabilis	Z,Ft,B	U,Sc	S,M	U	si
C. torquatus	Ft,Fh	Sc,C	S	R	t
Dryocopus lineatus	Z,Ft	Sc,C	S	F	t
Melanerpes cruentatus	Ft,Fh	C	S,M	C	t
Veniliornis passerinus	Z,C	Sc,C	S,M	F	t
V. affinis	Ft,Fh	Sc,C	M	F	t
Campephilus melanoleucus	Ft,Z	Sc	S	F	t
C. rubricollis	Ft,Fh	U,Sc	S	F	t
DENDROCOLAPTIDAE (14)					
Dendrocincla fuliginosa	Ft	U	S,M,A	F	t
D. merula	Ft,Fh	U,Sc	A	U	sp
Deconychura longicauda	Ft,Fh	Sc	S,M	U	t
Sittasomus griseicapillus	Ft	U,Sc	M	F	t
Glyphorhynchus spirurus	Ft,Fh	U,Sc	S,M	C	sp
Dendrexetastes rufigula	Ft,Fh	Sc,C	S,M	F	t
Xiphocolaptes promeropir	Fh	U,Sc	S,M	R	t
Dendrocolaptes certhia	Ft,Fh	Sc	S,M,A	F	t
D. picumnus	Ft,Fh	U,Sc	S,A	U	t
Xiphorhynchus picus	Z,C	Sc	S	R	t
X. spixii	Ft,Fh	U,Sc	M	C	sp
X. guttatus	Ft,Fh	Sc,C	S,M	C	t
Lepidocolaptes albolineatus	Ft,Fh	C	M	F	t
Campylorhamphus trochilir	Ft,B	U	S,M	R	si
FURNARIIDAE (19)					
Furnarius leucopus	Z,Ft,S	T	S	F	t
Synallaxis albescens	C	U	S	R(Ms?)	si
S. gujanensis	Z,C	U	S	F	t
S. rutilans	Ft,Fh	T,U	S	F	sp
Cranioleuca gutturata	Ft	Sc,C	M	U	si
Hyloctistes subulatus	Ft	Sc	S,M	U	t?
Ancistrops strigilatus	Ft,Fh	Sc,C	M	C	t

Habitats
- **Fh** Upland forest
- **Ft** Floodplain forest
- **Fs** Swamp forest
- **Fsm** Forest stream margins
- **Fo** Forest openings
- **Z** "Zabolo"
- **B** Bamboo
- **C** Clearing
- **R** River
- **Rm** River margins
- **S** Shores
- **M** Marsh
- **O** Overhead

Foraging Position
- **T** Terrestrial
- **U** Undergrowth
- **Sc** Subcanopy
- **C** Canopy
- **W** Water
- **A** Aerial

Sociality
- **S** Solitary or in pairs
- **G** Gregarious
- **M** Mixed-species flocks
- **A** Army ant followers

Abundance
- **C** Common
- **F** Fairly common
- **U** Uncommon
- **R** Rare
- **(M)** Migrant
- **(Mn)** Migrant from north
- **(Ms)** Migrant from south

Evidence
- **sp** Specimen
- **t** Tape
- **ph** Photo
- **si** ID by sight or sound
- ***** First record for La Paz
- **+** First record for Bolivia

	Habitats	Foraging	Sociality	Abundance	Evidence
Philydor erythrocercus	Fh,Ft	Sc	M	F	sp
P. pyrrhodes	Ft,Fh	Sc,U	S,M	U	t
P. rufus	Z	C	S,M	U	t
P. erythropterus	Ft,Fh	C	M	F	t
Automolus infuscatus	Ft,Fh	U	M	C	sp*
A. dorsalis	Ft,Z,B	Sc	S,M	R	si+
A. ochrolaemus	Ft,Fh	U	S,M	F	t
A. rufipileatus	Z,Ft	U	S,M	F	t
Xenops milleri	Ft	C	M	R	si+
X. rutilans	Ft,Fh	C	M	F	t
X. minutus	Ft,Fh	U,Sc	M	F	t
Sclerurus caudacutus	Ft,Fh	T	S	F	t*
FORMICARIIDAE (42)					
Cymbilaimus lineatus	Ft,Fh	Sc	S,M	F	t
Frederickena unduligera	Ft, Fh	U	S	R	t+
Taraba major	Z,C	U	S	U	t
Thamnophilus doliatus	C	U	S,M	U	t
T. aethiops	Ft,Fh	U	S	F	sp
T. schistaceus	Ft, Fh	Sc	M	C	sp
Pygiptila stellaris	Ft,Fh	Sc,C	M	F	t*
Thamnomanes ardesiacus	Ft,Fh	U	M	F	sp*
T. schistogynous	Ft,Z	U,Sc	M	U	t
Myrmotherula brachyura	Ft, Fh	Sc,C	M	C	t
M. sclateri	Ft,Fh	C	M	C	t
M. hauxwelli	Ft,Fh	U	S,M	F	sp*
M. leucophthalma	Ft,Fh	U	M	F	sp
M. axillaris	Ft,Fh	U,Sc	M	C	sp
M. longipennis	Fh	U,Sc	M	R?	t
M. menetriesii	Ft,Fh	Sc	M	C	sp
Dichrozona cincta	Fh	T	S	R?	t
Terenura humeralis	Ft, Fh	C	M	F	sp*
Cercomacra cinerascens	Ft,Fh	Sc	S,M	C	t
C. nigrescens	Z,B	U	S	U	t
C. serva	Fh,Ft,Fo	U	S	C	t
C. manu	Z,B	U	S	R	t*
Myrmoborus leucophrys	Ft	U	S	U	t
M. myotherinus	Ft,Fh	U	S,A	C	t
Hypocnemis cantator	Ft	U	S,M	C	t
Percnostola lophotes	Z,B	T,U	S	F	sp
Sclateria naevia	Ft,Fsm	T	S	U	t*
Myrmeciza hemimelaena	Ft,Fh	T,U	S	C	sp

APPENDIX 1

	Habitats	Foraging	Sociality	Abundance	Evidence
M. hyperythra	Fs,Ft	T,U	S	U	t
M. goeldii	Ft,B	U	S,A	U	t*
M. fortis	Fh	T,U	S,A	F	sp*
M. atrothorax	Z,C	U	S	C	t
Gymnopithys salvini	Ft,Fh	U	A	F	sp
Rhegmatorhina melanosticta	Fh	U	A	R	si
Hylophylax poecilinota	Ft,Fh	U	S,A	U	t
Phlegopsis nigromaculata	Ft	T,U	S,A	U	t
Chamaeza nobilis	Fh	T	S	R	si*
Formicarius colma	Fh	T	S	R	si
F. analis	Ft	T	S,A	C	t
Hylopezus berlepschi	Z,C	T	S	U	t
Myrmothera campanisona	Ft,Fh	T	S	C	t*
Conopophaga peruviana	Ft	U,T	S	R	t+
COTINGIDAE (5)					
Iodopleura isabellae	Ft,Fh	C	S	U	t*
Lipaugus vociferans	Ft,Fh	Sc,C	S	C	t
Cotinga maynana	Z,Ft	C	S	U	t*
Gymnoderus foetidus	Ft	C	S	R	si
Querula purpurata	Fh,Ft	C	G	C	t*
PIPRIDAE (7)					
Schiffornis turdinus	Fh	U	S	U	t
Piprites chloris	Ft,Fh	Sc,C	M	F	t
Tyranneutes stolzmanni	Ft,Fh	Sc	S	C	t
Machaeropterus pyrocephalus	Ft	Sc,C	S	F	sp
Pipra coronata	Fh	U,Sc	S	C	sp
P. fasciicauda	Ft	U,Sc	S	F	t
P. chloromeros	Ft,Fh	U,Sc	S	F	sp
TYRANNIDAE (66)					
Zimmerius gracilipes	Ft,Fh	C	S,M	C	t
Ornithion inerme	Ft,Fh	C	S,M	F	t
Camptostoma obsoletum	C,Z	C	S,M	U	t
Sublegatus obscurior	Z	Sc,C	S,M	R	si
Phaeomyias murina	C,Z	C	S,M	U	t
Tyrannulus elatus	Ft,Fh,C	Sc,C	S,M	F	t
Myiopagis gaimardii	Ft,Fh	C	M	C	t
M. caniceps	Ft	C	M	R	t
M. viridicata	Ft,Fh,Z	Sc,C	M	U(Ms?)	t
Elaenia spectabilis	Z	Sc,C	S,M	U(Ms)	si
Inezia inornata	C,Z	Sc,C	G,M	F(Ms)	si

Habitats	
Fh	Upland forest
Ft	Floodplain forest
Fs	Swamp forest
Fsm	Forest stream margins
Fo	Forest openings
Z	"Zabolo"
B	Bamboo
C	Clearing
R	River
Rm	River margins
S	Shores
M	Marsh
O	Overhead
Foraging Position	
T	Terrestrial
U	Undergrowth
Sc	Subcanopy
C	Canopy
W	Water
A	Aerial
Sociality	
S	Solitary or in pairs
G	Gregarious
M	Mixed-species flocks
A	Army ant followers
Abundance	
C	Common
F	Fairly common
U	Uncommon
R	Rare
(M)	Migrant
(Mn)	Migrant from north
(Ms)	Migrant from south
Evidence	
sp	Specimen
t	Tape
ph	Photo
si	ID by sight or sound
*	First record for La Paz
+	First record for Bolivia

	Habitats	Foraging	Sociality	Abundance	Evidence
Euscarthmus meloryphus	C	U	S	R(Ms?)	t
Mionectes olivaceus	Ft,Fh	Sc,C	S,M	F	sp+
M. oleagineus	Ft,Fh	Sc,U	M	U	t
M. macconnelli	Ft,Fh	U,Sc	M	R	sp
Leptopogon amaurocephalus	Ft	U,Sc	M,S	F	t
Corythopis torquata	Ft,Fh	T	S	F	t
Myiornis ecaudatus	Ft	Sc,C	S	F	t
Hemitriccus zosterops	Ft,Fh	Sc	S	F	t
Todirostrum latirostre	Z	U	S	F	t
T. maculatum	Z	Sc,C	S,M	F	si
T. chrysocrotaphum	Ft,Fh	C	S,M	F	t
Ramphotrigon megacephala	Z,B	Sc	S,M	F	sp
R. ruficauda	Ft,Fh	Sc	S	F	t
Rhynchocyclus olivaceus	Ft	U,Sc	M	U	t
Tolmomyias assimilis	Ft,Fh	C	M	F	t
T. poliocephalus	Ft,Z	Sc,C	S,M	U	t
T. flaviventris	Z	Sc,C	S,M	F	t
Platyrinchus coronatus	Ft	U,Sc	S	F	sp
Onychorhynchus coronatus	Ft,Fh,Fsm	U,Sc	S,M	U	sp
Terenotriccus erythrurus	Ft,Fh	U,Sc	S,M	U	t
Myiophobus fasciatus	C,Z	U	S,M	U	t
Contopus cinereus	Z	C,A	S	R(M?)	sp
Lathrotriccus euleri	Ft,Z,B	U,Sc	S	F	t
Cnemotriccus fuscatus	Z	U,Sc	S	U	t
Pyrocephalus rubinus	Rm,C	C,A	S	F(Ms)	si
Ochthoeca littoralis	Rm,S	T,A	S	F	t
Muscisaxicola fluviatilis	Rm,S	T	S	U	si
Hymenops perspicillata	M	U	S	R(Ms)	si
Satrapa icterophrys	Rm	Sc	S,M	R(Ms)	si
Attila cinnamomeus	Ft	Sc	S	R	t
A. bolivianus	Ft	Sc,C	S	U	t
A. spadiceus	Ft,Fh	Sc,C	S,M	F	t
Casiornis rufa	Ft	C	M	R(Ms)	si
Rhytipterna simplex	Ft,Fh	Sc,C	S,M	C	t
Laniocera hypopyrra	Ft,Fh	Sc,C	S,M	F	t
Sirystes sibilator	Ft,Fh	C,Sc	S,M	F	t
Myiarchus tuberculifer	Ft	C	S	R(M)	t
M. swainsoni	Z,Ft	C	M	U(Ms)	t?
M. ferox	Z,C	Sc,C	S,M	F	t
M. tyrannulus	Ft,Fh	C	S,M	F(Ms)	t
Pitangus lictor	M	U	S	R	t

APPENDIX 1

	Habitats	Foraging	Sociality	Abundance	Evidence
P. sulphuratus	Rm,C	U,Sc	S	F	t
Megarynchus pitangua	C,Z	Sc,C	S	F	t
Myiozetetes similis	C,Rm	U,Sc,C	S,G	C	t
M. granadensis	Z,C	Sc,C	S,G	F	t
M. luteiventris	Ft	C	S,G	U	t
Myiodynastes maculatus	Ft	Sc,C	M	U(Ms)	si
Empidonomus varius	Ft	C	M	R(Ms)	si
Tyrannus melancholicus	Rm,C	C,A	S	F	t
Pachyramphus polychopterus	Z,C	C	S,M	F	t
P. marginatus	Ft,Fh	C	M	F	t
P. minor	Ft,Fh	C	M	U	t
P. validus	Fh	C	M	R	si
Tityra cayana	Ft	C	S	F	si*
T. semifasciata	Ft,Fh	C	S	U	t
HIRUNDINIDAE (5)					
Tachycineta albiventer	R	A	S,G	C	t
Notiochelidon cyanoleuca	R	A	G	R(Ms?)	si
Atticora fasciata	R,C	A	G	C	t
Neochelidon tibialis	Fh	A	G	U	sp*
Stelgidopteryx ruficollis	R,C	A	G	C	t
CORVIDAE (1)					
Cyanocorax violaceus	Z,Ft	Sc,C	G,M	C	t*
TROGLODYTIDAE (7)					
Campylorhynchus turdinus	C,Z	Sc,C	S,M	F	t
Thryothorus genibarbis	Z,Ft	U	S	C	t
T. leucotis	C	U	S	U	t
Troglodytes aedon	C	U	S	U	t
Microcerculus marginatus	Ft,Fh	T	S	F	t
Cyphorhinus arada	Ft,Fh	T	S,M	F	t
Donacobius atricapillus	M,C	U	S	F	t
TURDIDAE (3)					
Turdus amaurochalinus	Z,C,Ft	T,Sc,C	S	F(Ms?)	t
T. lawrencii	Ft	T,Sc,C	S	R	t
T. albicollis	Ft,Fh	T,Sc	S	F	t
VIREONIDAE (4)					
Cyclarhis gujanensis	Z	C	S,M	R	t
Vireo olivaceus	Ft,Fh,Z	C	M	C(Ms)	t
Hylophilus hypoxanthus	Ft,Fh	C	M	C	t
H. ochraceiceps	Fh	U	M	R	t
ICTERIDAE (11)					

Habitats	
Fh	Upland forest
Ft	Floodplain forest
Fs	Swamp forest
Fsm	Forest stream margins
Fo	Forest openings
Z	"Zabolo"
B	Bamboo
C	Clearing
R	River
Rm	River margins
S	Shores
M	Marsh
O	Overhead
Foraging Position	
T	Terrestrial
U	Undergrowth
Sc	Subcanopy
C	Canopy
W	Water
A	Aerial
Sociality	
S	Solitary or in pairs
G	Gregarious
M	Mixed-species flocks
A	Army ant followers
Abundance	
C	Common
F	Fairly common
U	Uncommon
R	Rare
(M)	Migrant
(Mn)	Migrant from north
(Ms)	Migrant from south
Evidence	
sp	Specimen
t	Tape
ph	Photo
si	ID by sight or sound
*	First record for La Paz
+	First record for Bolivia

	Habitats	Foraging	Sociality	Abundance	Evidence
Molothrus bonariensis	Rm	T	G	R	si
Scaphidura oryzivora	Ft,Rm,S	T,C	S,G	F	t
Clypicterus oseryi	Ft	C	S, G	R	t+
Psarocolius decumanus	Ft,Fh	C	G,M	F	t
P. angustifrons	Ft,Z	Sc,C	G,M	C	t
Gymnostinops yuracares	Ft,Fs	C	G,M	F	t
Cacicus cela	Ft,Z	Sc,C	G,M	C	t
C. haemorrhous	Ft,Fh	Sc,C	S,G,M	U	t*
C. solitarius	Z	U,Sc	S	F	t
Icterus cayanensis	Ft,C	C	S,M	F	t
Leistes superciliaris	C	T,U	S,G	R(Ms)	si
PARULIDAE (2)					
Geothlypis aequinoctialis	M,C	U	S	F	t
Basileuterus rivularis	Ft,Fh,Fsm	T,U	S	F	t
COEREBIDAE (5)					
Cyanerpes caeruleus	Ft,Fh	C	S,G,M	C	si
Chlorophanes spiza	Ft,Fh	C	S,M	U	si
Dacnis cayana	Ft,Fh	C	M	C	si
D. lineata	Ft,Fh	C	M	C	t
D. flaviventer	Ft	C	S,M	U	t
TERSINIDAE (1)					
Tersina viridis	Z,Rm	C	G	U	t
THRAUPIDAE (25)					
Chlorophonia cyanea	Ft,Fh	C	M	R(M?)	t
Euphonia musica	Fh	C	S	R(M?)	si
E. xanthogaster	Ft,Fh	U,Sc,C	M	F	t
E. minuta	Ft	C	M	U	si
E. laniirostris	Z,C	Sc,C	S,M	U	t
E. rufiventris	Ft,Fh	Sc,C	S,M	C	t
E. chrysopasta	Ft	Sc,C	S,M	F	t
Tangara velia	Fh	C	M	U	t*
T. callophrys	Fh	C	M	U	t*
T. chilensis	Ft,Fh	Sc,C	G,M	C	t
T. schrankii	Ft,Fh	U,Sc,C	M	C	t
T. xanthogastra	Ft, Fh	C	M	R	si
T. nigrocincta	Ft,Fh	C	M	F	t
T. mexicana	Ft	C	M,G	F	t
Thraupis episcopus	Z,C	Sc,C	S,M	U	si
T. palmarum	Ft,C	C	S,M	F	t
Ramphocelus carbo	C,Z,Ft	U,Sc,C	G,M	C	t

	Habitats	Foraging	Sociality	Abundance	Evidence
R. nigrogularis	Ft,Fs	Sc,C	G,M	C	sp*
Habia rubica	Fh,Ft	U	G,M	C	t
Lanio versicolor	Ft,Fh	Sc,C	M	F	t
T. cristatus	Fh	C	M	U	si
Tachyphonus luctuosus	Z,Ft,Fh	Sc,C	M	C	t
Hemithraupis flavicollis	Ft, Fh	C	M	F	t
Conothraupis speculigera	Z	U,M	M	R	si+
Lamprospiza melanoleuca	Fh,Ft	C	M	F	sp*
Cissopis leveriana	Z,C	Sc,C	S,M	F	t
FRINGILLIDAE (11)					
Saltator maximus	Z,Ft,Fh	Sc,C	S,M	F	t
S. coerulescens	Z,C	U,Sc,C	S	F	t
Caryothraustes humeralis	Ft,Fh	C	M	R	si
Paroaria gularis	Rm	U,Sc,C	S,M	F	si
Cyanocompsa cyanoides	Ft	U	S	F	t
Volatinia jacariua	C,M	T,U	S,G	R	si
Sporophila caerulescens	C,M	T,U	G,M	C(Ms)	si
S. castaneiventris	C	U,C	S,M	U	si
Oryzoborus angolensis	C,M	U	S	R	si
Arremon taciturnus	Ft,Fh	T,U	S	F	t
Myospiza aurifrons	S,C	T,U	S	C	t

Habitats	
Fh	Upland forest
Ft	Floodplain forest
Fs	Swamp forest
Fsm	Forest stream margins
Fo	Forest openings
Z	"Zabolo"
B	Bamboo
C	Clearing
R	River
Rm	River margins
S	Shores
M	Marsh
O	Overhead
Foraging Position	
T	Terrestrial
U	Undergrowth
Sc	Subcanopy
C	Canopy
W	Water
A	Aerial
Sociality	
S	Solitary or in pairs
G	Gregarious
M	Mixed-species flocks
A	Army ant followers
Abundance	
C	Common
F	Fairly common
U	Uncommon
R	Rare
(M)	Migrant
(Mn)	Migrant from north
(Ms)	Migrant from south
Evidence	
sp	Specimen
t	Tape
ph	Photo
si	ID by sight or sound
*	First record for La Paz
+	First record for Bolivia

APPENDIX 2

Birds of the Lower Rio Heath, Bolivia/Peru

Theodore A. Parker, 1990

	Habitats	Foraging	Sociality	Abundance	Evidence
TINAMIDAE (8)					
Tinamus major	Ft	T	S	F	t*
T. guttatus	Fh,Ft	T	S	F	t
Crypturellus cinereus	Ft	T	S	C	t*
C. soui	Ft	T	S	F	t*
C. undulatus	Z,Ft	T	S	C	t*
C. bartletti	Ft	T	S	U	t*
C. parvirostris	Gf	T	S	U	t
Rhynchotus rufescens	P	T	S	F?	sp
PHALACROCORACIDAE (1)					
Phalacrocorax brasiliensis	R	W	S,G	U	si*
ANHINGIDAE (1)					
Anhinga anhinga	R	W	S	R	si*
ARDEIDAE (7)					
Ardea cocoi	S	W	S	U	si*
Egretta alba	S	W	S	U	si*
E. thula	S	W	S	U	si*
Butorides striatus	M	W	S	R	si*
Ardeola ibis	S	T	S,G	F	si*
Pilherodias pileatus	S,M	W	S	F	si*
Tigrisoma lineatum	M,Fsm	W	S	U	t*
CICONIIDAE (2)					
Mycteria americana	S	W	G,S	U	si*
Jabiru mycteria	S,P	W	S	R	si*
THRESKIORNITHIDAE (2)					
Mesembrinibis cayennensis	Fsm	T,W	S	U	si*
Ajaia ajaja	S	W	G	U/R	si*
ANHIMIDAE (1)					
Anhima cornuta	M,S	T	S	F	t*
ANATIDAE (2)					
Neochen jubata	S	T	S	U	si*
Cairina moschata	R,M	W	S	U	si*
CATHARTIDAE (5)					
Sarcoramphus papa	Ft	T	S	F	si*
Coragyps atratus	S,Z,Ft	T	G	U	si*
Cathartes aura	Z,Ft,Gf	T	S	U/R	si
C. burrovianus	P	T	S	F	ph
C. melambrotus	Z,Ft,Fh	T	S	F*	si
ACCIPITRIDAE (18)					

APPENDIX 2

	Habitats	Foraging	Sociality	Abundance	Evidence
Gampsonyx swainsonii	Rm,Z	A	S	R	si
Elanoides forflcatus	Ft	A	S,G	R	si*
Leptodon cayanensis	Ft,Fh	C,Sc	S	U	si*
Chondrohierax uncinatus	Ft	C	S	R	si*
Harpagus bidentatus	Fh	Sc	S	U	t*
Ictinia plumbea	Ft, Fh	A	S,G	F	si*
Accipiter bicolor	Ft	U,Sc	S	R	si
Buteo albicaudatus	P	T	S	U/R	si
B. magnirostris	Rm,Z,Gf	T,U,Sc	S	F	t*
Leucopternis kuhli	Fh	Sc	S	R	si
L. schistacea	Ft	T,Sc	S	U	t*
Busarellus nigricollis	Fs,A	T,W	S	U	t*
Buteogallus urubitinga	Rm,S	T,W	S	U	si*
Morphnus guianensis	Ft	Sc,C	S	R	si
Harpyia harpyja	Ft,Fh	C	S	R	si
Spizaetus ornatus	Ft	C	S	U	t*
S. tyrannus	Z,Ft	Sc,C	S	U	t*
Geranospiza caerulescens	Ft	T,U,Sc	S	U	si*
PANDIONIDAE (1)					
Pandion haliaetus	R	W	S	R(Mn)	si*
FALCONIDAE (9)					
Herpetotheres cachinnans	Ft,Rm	C,Sc,T	S	F	t
Micrastur ruficollis	Fh	U,Sc	S	R	si*
M. gilvicollis	Ft,Fh	U,Sc	S	U	t
Daptrius ater	Rm,S	T,C	S,G	U	t*
D. americanus	Ft,Fh	Sc,C	G	F	t*
Milvago chimachima	P	T	S	U	t
Polyborus plancus	P	T	S	R	t
Falco rufigularis	Rm,C	A	S	F	t*
F. femoralis	P	A	S	R	si
CRACIDAE (4)					
Ortalis motmot	Z,Ft,Gf	Sc,C	G	C	t*
Penelope jacquacu	Ft,Fh,Gf	C,Sc,T	S,G	F	t*
Aburria pipile	Ft,Z,Gf	C	S	C	si*
Mitu tuberosa	Ft,Gf	T	S	F	t
PHASIANIDAE (1)					
Odontophorus stellatus	Ft,Fh	T	G	U	t*
OPISTHOCOMIDAE (1)					
Opisthocomus hoazin	Rm	U,Sc,C	G	C	si
PSOPHIIDAE (1)					

Habitats	
A	Aguajales
Gf	Gallery forest
Lm	Lake margin
P	Pantanal-like grassland
Fh	Upland forest
Ft	Floodplain forest
Fsm	Forest stream margins
Fo	Forest openings
Z	"Zabolo"
B	Bamboo
C	Clearing
R	River
Rm	River margins
S	Shores
M	Marsh
O	Overhead
Foraging Position	
T	Terrestrial
U	Undergrowth
Sc	Subcanopy
C	Canopy
W	Water
A	AeriaL
Sociality	
S	Solitary or in pairs
G	Gregarious
M	Mixed-species flocks
A	Army ant followers
Abundance	
C	Common
F	Fairly common
U	Uncommon
R	Rare
(M)	Migrant
(Mn)	Migrant from north
(Ms)	Migrant from south
Evidence	
sp	Specimen
t	Tape
ph	Photo
si	Species ID by sight

Asterisked species were observed on both the Bolivian and Peruvian sides of the Rio Heath; those without * were noted in Peru only.

	Habitats	Foraging	Sociality	Abundance	Evidence
Psophia leucoptera	Fh	T	G	F	si
RALLIDAE (4)					
Aramides cajanea	Ft,Fsm	T	S	F	si
Porzana albicollis	P	T,U	S	C	sp
Laterallus exilis	M,C	T,U	S	F	t
Micropygia schomburgkii	P	T	S	F	sp
HELIORNITHIDAE (1)					
Heliornis fulica	R	W	S	U	si*
EURYPYGIDAE (1)					
Eurypyga helias	Ft,Fsm,S	T,W	S	U	si*
CHARADRIIDAE (2)					
Hoploxpterus cayanus	S	T	S	F	si*
Charadrius collaris	S	T	S	U	t*
SCOLOPACIDAE (5)					
Tringa solitaria	S	T,W	S	F(Mn)	t*
T. flavipes	S	T,W	S	F(Mn)	t*
T. melanoleuca	S	W	S,G	F(Mn)	t*
Actitis macularia	S	T	S	F(Mn)	t*
Calidris melanotos	S	T	G	F(Mn)	t*
LARIDAE (2)					
Phaetusa simplex	R	W,A	S	U	si*
Sterna superciliaris	R	W,A	S	F	si*
RHYNCHOPIDAE (1)					
Rynchops nigra	R	W	S	R	si*
COLUMBIDAE (9)					
Columba speciosa	Gf	C	S,G	F	t
C. cayennensis	Gf,Rm,Z	U,Sc,C	S,G	F	t*
C. subvinacea	Ft	C	S	U	t*
C. plumbea	Ft, Fh	C	S	C	t*
Columbina talpacoti	Z	T	S,G	F	t*
C. picui	Z,Rm	T	S,G	U(Ms?)	si*
Claravis pretiosa	Z,Ft	T	S	R	si*
Leptotila rufaxilla	Z	T	S	F	t*
Geotrygon montana	Ft,Fh	T	S	F	t
PSITTACIDAE (20)					
Ara ararauna	Ft,A	C	G	U	t*
A. macao	Ft, Fh	C	S,G	F	t*
A. chloroptera	Ft,Fh	C	S,G	F	t*
A. severa	Ft,Z,A	C	S,G	F	t*

APPENDIX 2

	Habitats	Foraging	Sociality	Abundance	Evidence
A. manilata	A	C	G	C	t*
A. couloni	Ft	C	G	R	si*
A. nobilis	Gf,A	C	G	C	sp
Aratinga leucophthalmus	Z,Ft,Gf	C	G	U	t*
A. weddellii	Z,Ft	C	G	C	t*
A. aurea	Gf	C	S,G	F	sp*
Pyrrhura rupicola	Ft,Fh	C	G	U	t*
Forpus sclateri	Ft, Fh	C	S	U	t*
Brotogeris cyanoptera	Ft,Fh,Z	C	G	C	t*
Nannopsittaca dachilleae	Ft	C	S	R	t*
Touit huetii	Ft	C	G	R	t*
Pionites leucogaster	Ft, Fh	C	G	C	t*
Pionopsitta barrabandi	Fh	Sc,C	S,G	F	t*
Pionus menstruus	Ft,Z,Gf	C	S,G	F	t*
Amazona ochrocephala	Ft,Z,Gf	C	S	F	t*
A. farinosa	Ft,Fh	C	S,G	F	t*
CUCULIDAE (7)					
Coccyzus melacoryphus	Z,Gf	C	S,M	U(Ms)	si
Piaya cayana	Z,Ft,Fh,Gf	Sc,C	S,M	F	t
P. minuta	Z	U	S	R	t
Crotophaga ani	P,Z	U,T	G	F	t
Tapera naevia	P	U	S	R	si
Dromococcyx phasianellus	Ft	T,U	S	R	t*
D. pavoninus	Ft	U	S	R	t*
STRIGIDAE (7)					
Otus choliba	Gf	Sc,C	S	U	t
O. watsonii	Ft, Fh	Sc,C	S	C	t*
Lophostrix cristata	Ft,Fh	Sc,C	S	U	t*
Pulsatrix perspicillata	Ft	Sc,C	S	U	t*
Glaucidium minutissimum	Ft,Fh	Sc	S	F	t*
G. brasilianum	Gf,Z	C,Sc	S	U	t
Ciccaba (virgata)	Ft,Fh	C,Sc	S	U	si
NYCTIBIIDAE (2)					
Nyctibius grandis	Ft	C,A	S	U	t
N. griseus	Gf	C,A	S	F	t
CAPRIMULGIDAE (10)					
Chordeiles rupestris	S	A	G	C	t*
Podager nacunda	P	A	G	R(Ms?)	si
Nyctidromus albicollis	Z	A	S	F	t*
Nyctiphrynus ocellatus	Ft	A,U	S	U	t*

Habitats	
A	Aguajales
Gf	Gallery forest
Lm	Lake margin
P	Pantanal-like grassland
Fh	Upland forest
Ft	Floodplain forest
Fsm	Forest stream margins
Fo	Forest openings
Z	"Zabolo"
B	Bamboo
C	Clearing
R	River
Rm	River margins
S	Shores
M	Marsh
O	Overhead
Foraging Position	
T	Terrestrial
U	Undergrowth
Sc	Subcanopy
C	Canopy
W	Water
A	AeriaL
Sociality	
S	Solitary or in pairs
G	Gregarious
M	Mixed-species flocks
A	Army ant followers
Abundance	
C	Common
F	Fairly common
U	Uncommon
R	Rare
(M)	Migrant
(Mn)	Migrant from north
(Ms)	Migrant from south
Evidence	
sp	Specimen
t	Tape
ph	Photo
si	Species ID by sight

Asterisked species were observed on both the Bolivian and Peruvian sides of the Rio Heath; those without * were noted in Peru only.

APPENDIX 2

	Habitats	Foraging	Sociality	Abundance	Evidence
mulgus rufus	Gf	A	S	R(Ms)	si
C. sericocaudatus	Gf	A	S	R(Ms?)	t
C. maculicaudus	P	A	S	C	sp
C. parvulus	Gf	A	S	U/F(Ms)	t
Hydropsalis climacocerca	Rm,S	A	S	F	t*
H. brasiliana	P	A	S	F(Ms?)	si
APODIDAE (6)					
Streptoprocne zonaris	O	A	G	U	si*
Chaetura cinereiventris	O,Ft	A	G,M	F(M?)	t*
C. egregia	O,Ft	A	G,M	U(Ms?)	t?*
C. brachyura	O,Gf	A	G,M	R	si*
Panyptila cayennensis	O,Ft	A	S,M	U	t*
Reinarda squamata	A,Ft	A	S,G	C	t*
TROCHILIDAE (14)					
Glaucis hirsuta	Z,Ft	U,Sc	S	U	t
Threnetes leucurus	Ft	U	S	?	sp
Phaethornis philippi	Fh	U	S	F	sp
P. hispidus	Z,Ft	U	S	F	t*
P. ruber	Ft,Fh	U	S	C	t
Eupetomena macroura	P	A,C	S	F	sp
Florisuga mellivora	Ft,Fh	Sc,C	S	U	sp
Anthracothorax nigricollis	Rm	C	S	U	si
Thalurania furcata	Ft,Fh	U,Sc	S	F	sp*
Hylocharis cyanus	Ft,Gf	U,Sc,C	S	F	t*
Polytmus guainumbi	P	U,Sc	S	U	sp
P. theresiae	P	U,Sc	S	U	sp
Amazilia lactea	Z	C	S	R	si
Heliomaster longirostris	Z,C	C	S	R	si*
TROGONIDAE (6)					
Pharomachrus pavoninus	Ft,Fh	Sc,C	S	U	t*
Trogon melanurus	Ft,Fh,Gf	Sc,C	S	F	t*
T. viridis	Ft,Fh,Gf	Sc,C	S,M	F	t*
T. collaris	Ft	U,Sc	S,M	F	t*
T. curucui	Ft,Z	Sc,C	S,M	U	t*
T. violaceus	Ft,Fh	Sc,C	S,M	U	t
ALCEDINIDAE (5)					
Ceryle torquata	Rm	W	S	F	t*
Chloroceryle amazona	Rm	W	S	F	t*
C. americana	Fsm,Rm	W	S	U	si*

	Habitats	Foraging	Sociality	Abundance	Evidence
C. inda	Fsm	W	S	U	si
C. aenea	Gf,Fsm	W	S	U	si
MOMOTIDAE (3)					
Electron platyrhynchum	Ft,Fh	Sc	S	C	t*
Baryphthengus martii	Ft	Sc	S	F	t*
Momotus momota	Ft,Fh	Sc	S	U	t
GALBULIDAE (4)					
Brachygalba albogularis	Z	A,C	S	R	si*
Galbula cyanescens	Ft,Gf	U,Sc	S,M	F	t*
G. dea	Ft	A,C	S,M	U	si
Jacamerops aurea	Ft,Fh	Sc,C	S	U	t*
BUCCONIDAE (8)					
Notharchus macrorhynchus	Ft,Fh	C	S	U	t
Bucco macrodactylus	Ft	Sc,C		U/R	sp
Nystalus chacuru	Gf	C	S	F	sp
N. striolatus	Ft	Sc,C	S	U	t*
Malacoptila semicincta	Fh	U	S	U	sp
Monasa nigrifrons	Z,Ft	Sc,C	G,M	C	t*
M. morphoeus	Fh	Sc,C	G,M	F	t
Chelidoptera tenebrosa	Rm,Z	A	S	C	t*
CAPITONIDAE (2)					
Capito niger	Ft,Fh	Sc,C	S,M	F	t*
Eubucco richardsoni	Ft	Sc,C	S,M	U	t*
RAMPHASTIDAE (8)					
Aulacorhynchus prasinus	Ft	C	S	U	t*
Pteroglossus castanotis	Ft,Z	C	G	F	t*
P. inscriptus	Ft,Z	Sc,C	G	R	t
P. mariae	Ft, Fh	C	G	F	t*
Pteroglossus beauharnaesii	Ft,Fh	C	G	U	t*
Ramphastos culminatus	Ft, Fh	C	S,G	C	t*
R. cuvieri	Ft,Fh	C	S,G	C	t*
R. toco	Gf	C	S	R	t
PICIDAE (15)					
Picumnus rufiventris	Z	U,Sc	S,M	R	sp
P. borbae	Ft	Sc,C	S,M	U	si
Chrysoptilus punctigula	Z	Sc,C	S	R	t*
Piculus chrysochloros	Ft	C	S,M	U	t?*
Celeus elegans	Ft	Sc	S	U	sp
C. grammicus	Fh	C	S,M	F	t*
C. flavus	Ft	Sc,C	S,G	U	t*

Habitats	
A	Aguajales
Gf	Gallery forest
Lm	Lake margin
P	Pantanal-like grassland
Fh	Upland forest
Ft	Floodplain forest
Fsm	Forest stream margins
Fo	Forest openings
Z	"Zabolo"
B	Bamboo
C	Clearing
R	River
Rm	River margins
S	Shores
M	Marsh
O	Overhead
Foraging Position	
T	Terrestrial
U	Undergrowth
Sc	Subcanopy
C	Canopy
W	Water
A	AeriaL
Sociality	
S	Solitary or in pairs
G	Gregarious
M	Mixed-species flocks
A	Army ant followers
Abundance	
C	Common
F	Fairly common
U	Uncommon
R	Rare
(M)	Migrant
(Mn)	Migrant from north
(Ms)	Migrant from south
Evidence	
sp	Specimen
t	Tape
ph	Photo
si	Species ID by sight

Asterisked species were observed on both the Bolivian and Peruvian sides of the Rio Heath; those without * were noted in Peru only.

	Habitats	Foraging	Sociality	Abundance	Evidence
C. torquatus	Ft	Sc,C	S	R	t*
Dryocopus lineatus	Z,Ft,Gf	Sc,C	S	F	t*
Melanerpes cruentatus	Ft,Fh,	C	S,M	C	t
Leuconerpes candidus	P,Gf	T,C	S	R	si
Veniliornis passerinus	Z	Sc,C	S,M	F	t*
V. affinis	Ft, Fh	Sc,C	M	F	t*
Campephilus melanoleucus	Ft,Z	Sc	S	F	t*
C. rubricollis	Fh	U,Sc	S	F	t
DENDROCOLAPTIDAE (14)					
Dendrocincla fuliginosa	Ft	U,Sc	S,M,A	F	t
D. merula	Fh	U,Sc	S,M,A	U/R	sp
Deconychura longicauda	Ft,Fh	Sc	S,M	U	t*
Sittasomus griseicapillus	Ft	U,Sc	M	F	t*
Glyphorynchus spirurus	Ft,Fh	U,Sc	S,M	U	sp
Dendrexetastes rufigula	Ft	Sc,C	S,M	F	t*
Dendrocolaptes certhia	Ft,Fh	Sc	S,M,A	F	t*
D. picumnus	Ft, Fh	U,Sc	S,A	U	t*
Xiphorhynchus picus	Z	Sc	S	U	t*
X. obsoletus	Ft	Sc	S,M	R	si*
X. spixii	Ft,Fh	U,Sc	M	C	sp
X. guttatus	Ft,Fh	Sc,C	S,M	C	sp
Lepidocolaptes albolineatus	Ft,Fh	C	M	U	t
Campylorhamphus trochilir	Ft,B	U	S,M	R	si
FURNARIIDAE (20)					
Furnarius leucopus	Z,Ft	T	S	U	t*
Synallaxis hypospodia	P	U	S	C	sp
S. albescens	P	U	S	U(Ms?)	t
S. gujanensis	Z	U	S	F	t
S. rutilans	Fh	T,U	S	F	sp
Cranioleuca gutturata	Ft	Sc,C	M	U	si
Thripophaga fusciceps	Ft	Sc	S,M	U	t*
Berlepschia rikeri	A	C	S	F	t
Ancistrops strigilatus	Ft,Fh	Sc,C	M	F	t*
Philydor erythrocercus	Fh, Ft	Sc	M	U	sp
P. pyrrhodes	Ft	Sc,U	S,M	U/R	t
P. rufus	Z	C	S,M	U	t*
P. erythropterus	Ft,Fh	C	M	F	t*
P. ruficaudatus	Ft	Sc	M	U	t*
Automolus infuscatus	Fh	U	M	F	sp
A. ochrolaemus	Ft	U	S,M	F	t

APPENDIX 2

	Habitats	Foraging	Sociality	Abundance	Evidence
A. rufipileatus	Z,Ft	U	S,M	F	t*
Xenops milleri	Ft	C	M	R	si
X. minutus	Ft,Fh	U,Sc	M	F	t
Sclerurus caudacutus	Fh	T	S	U	t
FORMICARIIDAE (44)					
Cymbilaimus lineatus	Ft,Fh	Sc	S,M	F	t
C. sanctaemariae	Ft,B	Sc	S,M	F	t*
Frederickena unduligera	Ft	U	S	R	sp
Taraba major	Z	U	S	U	t*
Thamnophilus doliatus	P,Z	U	S,M	F	t*
T. aethiops	Ft,Fh	U	S	F	sp
T. schistaceus	Ft,Fh	Sc	M	C	sp*
Pygiptila stellaris	Ft,Fh	Sc,C	M	F	t
Thamnomanes ardesiacus	Ft,Fh	U	M	U	sp
T. schistogynous	Ft,Z	U,Sc	M	F	t*
T. amazonicus	Gf	U,Sc	S,M	F	t
Myrmotherula brachyura	Ft, Fh	Sc,C	M	C	t*
M. sclateri	Ft,Fh	C	M	C	t*
M. surinamensis	Ft	Sc	S,M	U	t*
M. hauxwelli	Ft,Fh	U	S,M	F	sp
M. ornata	Ft,B	Sc	M	U	t*
M. leucophthalma	Ft,Fh	U	M	U	t?
M. axillaris	Ft, Fh	U,Sc	M	C	sp*
M. longipennis	Fh	U,Sc	M	R?	sp
M. menetriesii	Ft,Fh	Sc	M	C	sp*
Dichrozona cincta	Fh	T	S	R?	t
Formicivora rufa	P	U	S	F	sp
Terenura humeralis	Ft,Fh	C	M	F	t?
Drymophila devillei	Ft,B	Sc	S,M		t
Cercomacra cinerascens	Ft,Fh	Sc	S,M	F	t*
C. nigrescens	Z,B	U	S	U	t
C. manu	Ft,B	Sc	S	U	t
Myrmoborus leucophrys	Ft	U	S	F	t*
M. myotherinus	Ft,Fh	U	S,A	C	t
Hypocnemis cantator	Ft	U	S,M	F	t
Hypocnemoides maculicauda	Fsm	T,U	S	U	t
Percnostola lophotes	Z,B	T,U	S	F	t*
Myrmeciza hemimelaena	Ft,Fh	T,U	S	C	sp*
M. hyperythra	Ft	T,U	S	U	t*
M. goeldii	Ft,B	T,U	S,A	R	t
M. atrothorax	Z	T,U	S	U	t*

Habitats

A	Aguajales
Gf	Gallery forest
Lm	Lake margin
P	Pantanal-like grassland
Fh	Upland forest
Ft	Floodplain forest
Fsm	Forest stream margins
Fo	Forest openings
Z	"Zabolo"
B	Bamboo
C	Clearing
R	River
Rm	River margins
S	Shores
M	Marsh
O	Overhead

Foraging Position

T	Terrestrial
U	Undergrowth
Sc	Subcanopy
C	Canopy
W	Water
A	AeriaL

Sociality

S	Solitary or in pairs
G	Gregarious
M	Mixed-species flocks
A	Army ant followers

Abundance

C	Common
F	Fairly common
U	Uncommon
R	Rare
(M)	Migrant
(Mn)	Migrant from north
(Ms)	Migrant from south

Evidence

sp	Specimen
t	Tape
ph	Photo
si	Species ID by sight

Asterisked species were observed on both the Bolivian and Peruvian sides of the Rio Heath; those without * were noted in Peru only.

	Habitats	Foraging	Sociality	Abundance	Evidence
Gymnopithys salvini	Ft,Fh	U	A	F	sp
Hylophylax poecilinota	Ft, Fh	U	S,A	U	t
Phlegopsis nigromaculata	Ft	T,U	S,A	U	t*
Formicarius colma	Fh	T	S	F	t
F. analis	Ft	T	S,A	C	t*
Hylopezus berlepschi	Z	T	S	U	t
Myrmothera campanisona	Ft,Fh	T	S	U	t
Conopophaga peruviana	Ft	U,T	S	R	si
COTINGIDAE (5)					
Iodopleura isabellae	Ft	C	S	U/R	si
Lipaugus vociferans	Ft, Fh	Sc,C	S	C	t*
Cotinga maynana	Z,Ft	C	S	F	si*
Gymnoderus foetidus	Ft	C	S,G	F	si*
Querula purpurata	Fh,Ft	C	G	U	t
PIPRIDAE (12)					
Schiffornis major	Ft	U	S	U	t*
S. turdinus	Fh	U	S	U	sp
Piprites chloris	Ft,Fh	Sc,C	M	F	t*
Xenopipo atronitens	Gf	U,Sc	S	R?	sp
Heterocercus linteatus	Ft?	Sc	S	R	si
Tyranneutes stolzmanni	Ft, Fh	Sc	S	C	t*
Manacus manacus	Ft,Gf	U	S	U	sp
Machaeropterus pyrocephalus	Ft	Sc,C	S	F	sp
Pipra coronata	Fh	U,Sc	S	R	sp
P. fasciicauda	Ft	U,Sc	S	F	t
P. rubrocapilla	Fh	U,Sc	S	F	sp
P. chloromeros	Ft	U,Sc	S	U	sp
TYRANNIDAE (65)					
Zimmerius gracilipes	Ft,Fh	C	S,M	F	t*
Ornithion inerme	Ft, Fh	C	S,M	F	t*
Camptostoma obsoletum	Z,Gf	C	S,M	U	t
Sublegatus obscurior	Z	Sc,C	S,M	R	si*
Phaeomyias murina	Z	C	S,M	U	t*
Tyrannulus elatus	Ft, Fh	C	S,M	F	t*
Myiopagis gaimardii	Ft,Fh	C	M	C	t*
M. caniceps	Ft	C	M	U	t*
M. viridicata	Ft,Fh,Z	Sc,C	M	U(Ms?)	t*
Elaenia flavogaster	Gf	C	C	sp	
E. spectabilis	Z	Sc,C	S,M	U(Ms)	si*
E. parvirostris	Z	C	S,M	U(Ms)	si*

APPENDIX 2

	Habitats	Foraging	Sociality	Abundance	Evidence
E. chiriquensis	Gf	C	S	U	sp
Inezia inornata	Z,Gf	Sc,C	G,M	F(Ms?)	si*
Euscarthmus meloryphus	Z,Gf	U	S	R(Ms?)	t
Mionectes oleagineus	Ft, Fh	Sc,U	M	U	sp
M. macconnelli	Ft,Fh	U,Sc	M	R	sp
Leptopogon amaurocephalus	Ft	U, Sc	M,S	F	t*
Corythopis torquata	Ft,Fh	T	S	F	t
Myiornis ecaudatus	Ft	Sc,C	S	F	t
Hemitriccus zosterops	Ft,Fh	Sc	S	F	t
H. iohannis	Z,Ft	Sc	S	U	t*
H. striaticollis	Gf	Sc,C	S	F	sp
Todirostrum latirostre	Z	U	S	F	t*
T. maculatum	Z	Sc,C	S,M	F	t*
T. chrysocrotaphum	Ft, Fh	C	S,M	F	t*
Ramphotrigon ruficauda	Ft,Fh	Sc	S	F	t*
Tolmomyias sulphurescens	Ft	C	M	U/R	si*
T. assimilis	Ft,Fh	C	M	F	t
T. poliocephalus	Ft,Z	Sc,C	S,M	U	t*
T. flaviventris	Z	Sc,C	S,M	F	t*
Platyrinchus coronatus	Ft	U,Sc	S	F	sp*
Onychorhynchus coronatus	Ft,Fh,Fsm	U,Sc	S,M	U	si
Terenotriccus erythrurus	Ft,Fh	U,Sc	S,M	U	t*
Myiophobus fasciatus	Gf,Z	U	S,M	F	t*
Lathrotriccus euleri	Ft,Z,B	U,Sc	S	F	t
Cnemotriccus fuscatus	Gf,Z	U,Sc	S	U	t*
Pyrocephalus rubinus	Rm,Z,Gf	C,A	S	F(Ms)	si*
Ochthoeca littoralis	Rm,S	T,A	S	F	t*
Muscisaxicola fluviatilis	Rm,S	T	S	U	si*
Fluvicola pica	Rm,S	T	S	R(Ms)	si*
Satrapa icterophrys	Rm,Z	Sc,C	S,M	R(Ms)	si*
Attila bolivianus	Ft	Sc,C	S,M	F	t*
A. spadiceus	Ft, Fh	Sc,C	S,M	F	t*
Rhytipterna simplex	Ft,Fh	Sc,C	S,M	C	t*
Sirystes sibilator	Ft	C,Sc	S,M	F	t*
Myiarchus swainsoni	Gf,Z,Ft	C	M	U(Ms)	t?
M. ferox	Z,C	Sc,C	S,M	F	t*
M. tyrannulus	Ft,Fh	C	S,M	F(Ms)	t
Pitangus sulphuratus	Rm,Z	U,Sc	S	F	t*
Megarynchus pitangua	C,Z	Sc,C	S	F	t*
Myiozetetes cayanensis	Lm	U,C	S	U	si*
M. similis	Rm,Z	C	S,G	C	t*

Habitats

A	Aguajales
Gf	Gallery forest
Lm	Lake margin
P	Pantanal-like grassland
Fh	Upland forest
Ft	Floodplain forest
Fsm	Forest stream margins
Fo	Forest openings
Z	"Zabolo"
B	Bamboo
C	Clearing
R	River
Rm	River margins
S	Shores
M	Marsh
O	Overhead

Foraging Position

T	Terrestrial
U	Undergrowth
Sc	Subcanopy
C	Canopy
W	Water
A	AeriaL

Sociality

S	Solitary or in pairs
G	Gregarious
M	Mixed-species flocks
A	Army ant followers

Abundance

C	Common
F	Fairly common
U	Uncommon
R	Rare
(M)	Migrant
(Mn)	Migrant from north
(Ms)	Migrant from south

Evidence

sp	Specimen
t	Tape
ph	Photo
si	Species ID by sight

Asterisked species were observed on both the Bolivian and Peruvian sides of the Rio Heath; those without * were noted in Peru only.

	Habitats	Foraging	Sociality	Abundance	Evidence
M. granadensis	Z,Ft	C	S,G	F	t*
M. luteiventris	Ft	C	S,G	U	t*
Myiodynastes maculatus	Ft	Sc,C	M	U(Ms)	si
Legatus leucophaius	Ft,Gf	A,C	S	U	t*
Empidonomus varius	Ft	C	M	R(Ms)	si*
Tyrannopsis sulphurea	A	C	S	U	sp
Tyrannus albogularis	A	A,C	S	U	t
T. melancholicus	Rm,Z,Gf	A,C	S	F	t*
Pachyramphus polychopterus	Z,Gf	C	S,M	F	t*
P. marginatus	Ft,Fh	C	M	F	t*
Tityra cayana	Ft	C	S	F	si*
T. inquisitor	Ft	C	S	U	si*
HIRUNDINIDAE (6)					
Phaeoprogne tapera	Rm	A	S,G	F	t*
Progne chalybea	Rm	A	S,G	U	si*
Tachycineta albiventer	R	A	S,G	C	t
Notiochelidon cyanoleuca	R,C	A	G	R(Ms?)	si
Atticora fasciata	R,C	A	G	C	t
Stelgidopteryx ruficollis	R,C	A	G	C	t
CORVIDAE (1)					
Cyanocorax violaceus	Z,Ft	Sc,C	G,M	C	t*
TROGLODYTIDAE (7)					
Campylorhynchus turdinus	C,Z	Sc,C	S,M	F	t*
Thryothorus genibarbis	Z,Ft,B	U	S	C	t*
T. leucotis	Z	U	S	U	t*
Troglodytes aedon	C,Rm	U	S	U	t
Microcerculus marginatus	Ft,Fh	T	S	F	t*
Cyphorhinus arada	Ft,Fh	T	S,M	F	t
Donacobius atricapillus	Lm	U	S	F	t*
TURDIDAE (6)					
Turdus leucomelas	Gf	T,C	S	U(Ms?)	sp
T. amaurochalinus	Z,Gf,Ft	T,C	S	F(Ms?)	t*
T. ignobilis	Z	T,C	S,G	F?	si*
T. lawrencii	Ft	T,C	S	F	t*
T. hauxwelli	Ft	T	S	U	t*
T. albicollis	Ft,Fh	T,Sc	S	F	t
MOTACILLIDAE (1)					
Anthus lutescens	P	T	S	U/R	sp?
VIREONIDAE (4)					
Cyclarhis gujanensis	Z	C	S,M	R	t

APPENDIX 2

	Habitats	Foraging	Sociality	Abundance	Evidence
Vireo olivaceus	Ft,Fh,Z,Gf	C	G,M	C(Ms)	t*
Hylophilus hypoxanthus	Ft,Fh	C	M	C	t*
Hylophilus sp.	Ft	C	M	U?	t
ICTERIDAE (8)					
Scaphidura oryzivora	Rm,S,P	T,C	S,G	F	t*
Psarocolius decumanus	Ft, Fh	C	G,M	F	t*
P. angustifrons	Ft,Z	Sc,C	G,M	F	t*
Gymnoscinops yuracares	Ft,Fs	C	G,M	F	t*
Cacicus cela	Ft,Z	Sc,C	G,M	C	t*
C. solitarius	Z	U,Sc	S	F	t*
Gnorimopsar chopi	P,A,Gf	C	G	F	sp
Icterus cayanensis	Ft,C	C	S,M	U	t
PARULIDAE (1)					
Geothlypis aequinoctialis	M,Z	U	S	U	t*
COEREBIDAE (6)					
Coereba flaveola	Gf	C	S	U	sp
Cyanerpes caeruleus	Ft,Fh,Gf	C	S,G,M	U	si
Chlorophanes spiza	Ft,Fh	C	S,M	U	si
Dacnis cayana	Ft, Fh	C	M	F	si
D. lineata	Ft,Fh	C	M	C	si*
D. flaviventer	Ft	C	S,M	U	t
TERSINIDAE (1)					
Tersina viridis	Z,Rm	C	G	U	t*
THRAUPIDAE (20)					
Chlorophonia cyanea	Ft,Fh	C	M	R(M?)	t
Euphonia minuta	Ft	C	M	U	si
E. chlorotica	Gf,Ft	C	S	F	t
E. laniirostris	Z	Sc,C	S,M	U	t
E. rufiventris	Ft	Sc,C	S,M	F	t*
E. chrysopasta	Ft	Sc,C	S,M	F	t*
Tangara velia	Fh	C	M	U	t
T. chilensis	Ft,Fh	Sc,C	G,M	C	t*
T. schrankii	Fh	U,Sc,C	M	C	t
T. mexicana	Ft	C	M,G	C	t*
Thraupis episcopus	Z	Sc,C	S,M	U	si*
T. palmarum	Ft,Gf	C	S,M	F	t*
Ramphocelus carbo	C,Z,Ft	U,Sc,C	G,M	C	t*
Habia rubica	Fh,Ft	U	G,M	U	t
Tachyphonus cristatus	Fh	C	M	U	sp
Tachyphonus luctuosus	Z,Ft,Fh	Sc,C	M	F	t*

Habitats	
A	Aguajales
Gf	Gallery forest
Lm	Lake margin
P	Pantanal-like grassland
Fh	Upland forest
Ft	Floodplain forest
Fsm	Forest stream margins
Fo	Forest openings
Z	"Zabolo"
B	Bamboo
C	Clearing
R	River
Rm	River margins
S	Shores
M	Marsh
O	Overhead
Foraging Position	
T	Terrestrial
U	Undergrowth
Sc	Subcanopy
C	Canopy
W	Water
A	AeriaL
Sociality	
S	Solitary or in pairs
G	Gregarious
M	Mixed-species flocks
A	Army ant followers
Abundance	
C	Common
F	Fairly common
U	Uncommon
R	Rare
(M)	Migrant
(Mn)	Migrant from north
(Ms)	Migrant from south
Evidence	
sp	Specimen
t	Tape
ph	Photo
si	Species ID by sight

Asterisked species were observed on both the Bolivian and Peruvian sides of the Rio Heath; those without * were noted in Peru only.

	Habitats	Foraging	Sociality	Abundance	Evidence
Lanio versicolor	Fh	C,Sc	M	U	sp
Hemithraupis flavicollis	Ft,Fh	C	M	F	t
Thlypopsis sordida	Z	U,C	M	R(Ms?)	si
Schistochlamys melanopis	P,Gf	U,C	S,G		sp
FRINGILLIDAE (13)					
Saltator maximus	Z,Ft,Fh	Sc,C	S,M	F	t*
S. coerulescens	Z,C	U,C	S	F	t
Paroaria gularis	Rm	U,C	S,M	F	si*
Cyanocompsa cyanoides	Ft	U	S	F	t*
Volatinia jacarina	P,M	T,U	S,G	U	si
Sporophila plumbea	P	U	S	U	sp
S. caerulescens	P,M	T,U	G,M	C(Ms)	si*
S. castaneiventris	P,M	U,C	S,M	U	si*
Oryzoborus angolensis	M	U	S	R	si
Myospiza humeralis	P	T	S	U	sp
M. aurifrons	S,C	T,U	S	C	t*
Emberizoides herbicola	P	U	S	F	sp
Coryphaspiza melanotis	P	U	S	F	sp

Birds of Ixiamas Area

Theodore A. Parker, 1990

	Habitats	Foraging	Sociality	Abundance	Evidence
TINAMIDAE (3)					
Crypturellus undulatus	Gf	T	S	C	t
C. parvirostris	Gf,Gw	T	S	F	t
Rhynchotus rufescens	P	T	S	F?	si
ARDEIDAE (7)					
Tigrisoma lineatum	M	W	S	F	t
*Syrigma sibilatrix**	M	T,W	S	R	si
Pilherodias pileatus	M	W	S	U	si
Ardea cocoi	M	W	S	U	si
Egretta alba	M	W	S	U	si
Bubulcus ibis	C,P	T	S,G	F	si
Nycticorax nycticorax	M	T	S,G	R	si
CICONIIDAE (1)					
Mycteria americana	M	W	G,S	U	si
THRESKIORNITHIDAE (2)					
*Theristicus caudatus**	P	T	S	U	t
Mesembrinibis cayennensis	Fsm,M	T,W	S	U	t
ANHIMIDAE (1)					
*Chauna torquata**	P,M	T	S	U	si
ANATIDAE (1)					
Cairina moschata	M	W	S	U	si
CATHARTIDAE (3)					
Coragyps atratus	C,Gf	T	G	F	si
Cathartes aura	C,Gf	T	S	F	si
*C. burrovianus**	P	T	S	F	si
ACCIPITRIDAE (7)					
Elanoides forficatus	Gf	C,A	S,G	R	si
*Gampsonyx swainsonii**	Gw	A	S	R	si
*Elanus caeruleus**	Gw	A,T	S	U	si
*Circus buffoni**	P	T	S	R?	si
*Buteogallus meridionalis**	P,Gw	T	S	U	si
Buteo magnirostris	Gf,C	T,Sc	S	C	t
*B. albicaudatus**	Gw	T	S	F	t
FALCONIDAE (3)					
*Polyborus plancus**	P	T	S	R	t
Herpetotheres cachinnans	Gf	T,C	S	F	t
Falco femoralis	P,Gw	A,T	S	R	si
RALLIDAE (4)					

Habitats	
A	Aguajales
Gf	Gallery forest
Gw	Wooded grasslands
P	Pantanal-like grasslands
Fsm	Forest stream margins
Fo	Forest openings
Z	"Zabolo"
B	Bamboo
C	Clearing
M	Marsh
O	Overhead
Foraging Position	
T	Terrestrial
U	Undergrowth
Sc	Subcanopy
C	Canopy
W	Water
A	AeriaL
Sociality	
S	Solitary or in pairs
G	Gregarious
M	Mixed-species flocks
A	Army ant followers
Abundance	
C	Common
F	Fairly common
U	Uncommon
R	Rare
(M)	Migrant
(Mn)	Migrant from north
(Ms)	Migrant from south
Evidence	
sp	Specimen
t	Tape
ph	Photo
si	Species ID by sight
*	Species new to La Paz department

	Habitats	Foraging	Sociality	Abundance	Evidence
Aramides cajanea	Gf,Fsm	T	S	F	si
*Porzana albicollis**	P	T,U	S	C	t
*Laterallus exilis**	M,P	T,U	S	F	t
*Micropygia schomburgkii**	P,Gw	T	S	F	t
ARAMIDAE (1)					
Aramus guarauna	M	W	S	U	t
CHARADRIIDAE (1)					
*Vanellus chilensis**	P	T	S,G	F	t
SCOLOPACIDAE (1)					
Gallinago paraguaiae	M,P	T	S	F	t
COLUMBIDAE (5)					
Columba cayennensis	Gf	U,Sc,C	S,G	C	t
Columbina talpacoti	Gw,Gf	T	S,G	F	si
C. picui	Gw,Gf	T	S,G	U(Ms?)	si
Claravis pretiosa	Gf	T	S	U	si
Leptotila rufaxilla	Gf	T	S	F	t
PSITTACIDAE (8)					
Ara ararauna	Gf,A	C	G	U	t
A. chloroptera	Gf	C	S,G	F	t
A. severa	Gf,A	C	S,G	F	t
A. manilata	A	C	G	F	t
Aratinga leucophthalmus	Gf	C	G	C	t
A. aurea	Gw,Gf	C	S,G	F	t
Pionus menstruus	Gf	C	S,G	F	t
Amazona ochrocephala	Gf	C	S	F	t
CUCULIDAE (2)					
Piaya cayana	Gf	Sc,C	S,M	F	t
Crotophaga ani	Gw	U,T	G	C	t
TYTONIDAE (1)					
Tyto alba	P,Wg	T	S	U?	si
STRIGIDAE (3)					
Otus choliba	Gf,Gw	Sc,C	S	F	t
Pulsatrix perspicillata	Gf	Sc,C	S	U	si
Glaucidium brasilianum	Gf	C,Sc	S	F	t
NYCTIBIIDAE (1)					
*Nyctibius griseus**	Gf	A,C	S	F	t
CAPRIMULGIDAE (1)					
Nyctidromus albicollis	Gf,Gw	A	S	C	t
APODIDAE (2)					

APPENDIX 3

	Habitats	Foraging	Sociality	Abundance	Evidence
Streptoprocne zonaris	O	A	G	U	si
*Tachornis squamata**	A,Gw	A	S,G	F	si
TROCHILIDAE (4)					
Phaethornis hispidus	Gf	U	S	F	t
Hylocharis cyanus	Gf	U,C	S	F	si
Polytmus guainumbi	P,Gw	U	S	C	t
Calliphlox amethystina	Gf	C	S	U	si
TROGONIDAE (1)					
Trogon viridis	Gf	C	S	F	t
ALCEDINIDAE (2)					
Ceryle torquata	M	W	S	U	t
Chloroceryle amazona	M	W	S	U	si
BUCCONIDAE (2)					
Nystalus chacuru	Gw	C	S	F	t
Monasa nigrifrons	Gf	Sc,C	G	C	t
RAMPHASTIDAE (3)					
Pteroglossus castanotis	Gf	C	G	F	t
Ramphastos culminatus	Gf	C	S,G	U	t
*R. toco**	Gf,Gw	C	S	U	si
PICIDAE (2)					
Dryocopus lineatus	Gf	Sc,C	S	F	si
*Leuconerpes candidus**	Gw,Gf	T,C	S	U	t
DENDROCOLAPTIDAE (2)					
Riphorhynchus guttatus	Gf	Sc,C	S,M	F	si
*Lepidocolaptes angustirostris**	Gw	U,C	S,M	U	t
FURNARIIDAE (8)					
Purnarius leucopus	Gf,Fsm	T	S	U	t
*Synallaxis frontalis**	Gf	U	S	R?	t
*S. hypospodia**	P	U	S	C	t
*S. albescens**	P	U	S	F	t
S. gujanensis	Gf	U	S	F	t
*Certhiaxis cinnamomea**	M	U	S	U	si
*Phacellodomus ruber**	Gw	U	S	F	t
*Berlepschia rikeri**	Gf	C	S	U	si
FORMICARIIDAE (2)					
Thamnophilus doliatus	Gf,Gw	U	S	F	t
*Formicivora rufa**	Gw	U	S	F	t
TYRANNIDAE (20)					
Camptostoma obsoletum	Gf,Gw	C	S,M	U	t

Habitats	
A	Aguajales
Gf	Gallery forest
Gw	Wooded grasslands
P	Pantanal-like grasslands
Fsm	Forest stream margins
Fo	Forest openings
Z	"Zabolo"
B	Bamboo
C	Clearing
M	Marsh
O	Overhead
Foraging Position	
T	Terrestrial
U	Undergrowth
Sc	Subcanopy
C	Canopy
W	Water
A	Aerial
Sociality	
S	Solitary or in pairs
G	Gregarious
M	Mixed-species flocks
A	Army ant followers
Abundance	
C	Common
F	Fairly common
U	Uncommon
R	Rare
(M)	Migrant
(Mn)	Migrant from north
(Ms)	Migrant from south
Evidence	
sp	Specimen
t	Tape
ph	Photo
si	Species ID by sight
*	Species new to La Paz department

	Habitats	Foraging	Sociality	Abundance	Evidence
Elaenia flavogaster	Gw,Gf	C	S	C	t
Inezia inornata	Gf	Sc,C	G,M	F(Ms?)	t
*Culicivora caudata**	P	U	S	R	si
*Euscarthmus meloryphus**	Gf,Gw	U	S	R(Ms?)	t
Hemitriccus striaticollis	Gf	Sc,C	S	F	t
Tolmomyias poliocephalus	Gf	C	S,M	U	t
Myiophobus fasciatus	Gf,Gw	U,Sc	S	F	t
Cnemotriccus fuscatus	Gf	U,Sc	S	F	t
Pyrocephalus rubinus	Gw,Gf	C,A	S	C(Ms)	t
*Alectrurus tricolor**	P	U	S	U	si
*Gubernetes yetapa**	P,Gw	C,T	S	U	t
Myiarchus swainsoni	Gf	C	M	U(Ms)	t?
M. ferox	Gf	Sc,C	S,M	U	t
M. tyrannulus	Gf	C	S,M	F(Ms)	t
Pitangus sulphuratus	Gf,Gw,M	U,Sc	S	F	t
Megarynchus pitangua	Gf	C	S	F	t
*Tyrannus albogularis**	A,Gf	A,C	S	U	si
T. melancholicus	Gf,Gw	A,C	S	F	t
Pachyramphus polychopterus	Gf	C	S,M	F	t
TROGLODYTIDAE (4)					
Campylorhynchus turdinus	Gf	Sc,C	S,M	F	t
Cistothorus platensis	P	U	S	U	si
Thryothorus (guarayanus?)	Gf	U	S	U	t
Troglodytes aedon	Gw	U	S	U	t
TURDIDAE (1)					
Turdus amaurochalinus	Gf	T,C	S	C(Ms?)	si
CORVIDAE (1)					
Cyanocorax cyanomelas	Gf,Gw	Sc,C	G	C	t
VIREONIDAE (2)					
Cyclarhis gujanensis	Gf	C	S,M	F	t
Vireo olivaceus	Gf	C	S,M	F(Ms)	si
MOTACILLIDAE (1)					
*Anthus lutescens**	P	T	S	F	t
EMBERIZINAE (10)					
Ammodramus humeralis	P,Gw	T,U	S	F	t
Sicalis sp.	P	T	S,G	R	si
Emberizoides herbicola	Gw,P	T,U	S	F	t
*Sporophila plumbea**	P,Gw	U	S,G	U	t
*S. collaris**	M	U	S,G	U	si
*S. hypochroma**	P	U	G	C	ph

	Habitats	Foraging	Sociality	Abundance	Evidence
*S. ruficollis**	P	U	G	R	si
*Sporophila** sp.	P	U	G	R	si
Oryzoborus angolensis	Gw	U	S	U	si
*Coryphaspiza melanotis**	P,Gw	U	S	U	t
CARDINALINAE (1)					
Saltator coerulescens	Gf	U,Sc	S	U	t
THRAUPINAE (4)					
Schistochlamys melanopis	Gw,Gf	U,C	S,G	C	t
Ramphocelus carbo	Gf	U,C	G	C	t
Thraupis episcopus	Gf,Gw	C	S,M	F	t
Euphonia chlorotica	Gf,Gw	C	S	F	t
*Tangara cayana**	Gf,Gw	C	S	F	t
Coereba flaveola	Gf	C	S	U	t
PARULIDAE (1)					
Geothlypis aequinoctialis	M	U	S	U	t
ICTERIDAE (5)					
Psarocolius decumanus	Gf	C	G,M	C	si
Cacicus cela	Gf	C	G,M	C	si
Leistes superciliaris	P,M	T,U	G	F	si
*Gnorimopsar chopi**	Gw,A	T,C	G	F	t

Habitats	
A	Aguajales
Gf	Gallery forest
Gw	Wooded grasslands
P	Pantanal-like grasslands
Fsm	Forest stream margins
Fo	Forest openings
Z	"Zabolo"
B	Bamboo
C	Clearing
M	Marsh
O	Overhead
Foraging Position	
T	Terrestrial
U	Undergrowth
Sc	Subcanopy
C	Canopy
W	Water
A	AeriaL
Sociality	
S	Solitary or in pairs
G	Gregarious
M	Mixed-species flocks
A	Army ant followers
Abundance	
C	Common
F	Fairly common
U	Uncommon
R	Rare
(M)	Migrant
(Mn)	Migrant from north
(Ms)	Migrant from south
Evidence	
sp	Specimen
t	Tape
ph	Photo
si	Species ID by sight
*	Species new to La Paz department

APPENDIX 4

Birds of Calabatea

Theodore A. Parker, 1990

	Habitats	Foraging	Sociality	Abundance	Evidence
TINAMIDAE (3)					
Tinamus tao	Fh	T	S	F	t
Nothocercus nigrocapillus	Fh	T	S	F	t
Crypturellus obsoletus	Fh	T	S	C	t
CATHARTIDAE (1)					
Cathartes aura	Fh	T	S	U	si
ACCIPITRIDAE (3)					
Buteo magnirostris	Fe	T,U,Sc	S	F	si
B. brachyurus	Fh	A,C	S	R	si
B. polyosoma	Fe,Sg	T	S	U	si
FALCONIDAE (1)					
Micrastur ruficollis	Fh	U,Sc	S	R	t
CRACIDAE (1)					
Penelope jacquacu	Fh	C,Sc,T	S,G	F	t
PHASIANIDAE (1)					
Odontophorus speciosus	Fh	T	G	U	t
COLUMBIDAE (3)					
Columba plumbea	Fh	C	S	C	t
Claravis mondetoura	Fh	T	S	R	si
Leptotila verreauxi	Sg	T	S	C	si
PSITTACIDAE (3)					
Ara militaris	Fh	C	G	U	t
Pionus sordidus	Fh	C	S,G	U	t
Amazona mercenaria	Fh	C	S,G	C	t
CUCULIDAE (2)					
Piaya cayana	Fh,Sg	Sc,C	S,M	F	t
Crotophaga ani	Sg	U,T	G	F	t
STRIGIDAE (2)					
Glaucidium jardinii	Fh	Sc,C	S	U	t
Ciccaba (huhula)	Fh	C	S	U	t
STEATORNITHIDAE (1)					
Steatornis caripensis	Fh	C	S	U	si
APODIDAE (3)					
Streptoprocne zonaris	Fh,Sg	A	G	F	si
Chaetura cinereiventris	Fh	A	S,G	F	t
Cypseloides rutilus	Fh,Sg	A	G	F	t
TROCHILIDAE (10)					
Phaethornis superciliosus	Fh	U	S	F	t

APPENDIX 4

	Habitats	Foraging	Sociality	Abundance	Evidence
Colibri delphinae	Fh	C	S	U	t
C. coruscans	Sg	U,C	S	F	si
Adelomyia melanogenys	Fh,Sg	U	S	F	t
Heliodoxa leadbeateri	Fh	U,Sc	S	F	t
Coeligena coeligena	Fh	U,Sc	S	F	si
Haplophaedia aureliae	Fh	U,Sc	S	U	si
Ocreatus underwoodii	Fe	U,Sc	S	F	si
Aglaiocercus kingi	Fh	C	S	U	si
Heliothryx aurita	Fh	C	S	R	si
Acestrura mulsant	Sg	U,C	S	U	si
TROGONIDAE (4)					
Pharomachrus auriceps	Fh	C	S	U	t
Trogon collaris	Fh	Sc	S	U	t
T. personatus	Fh	Sc	S	U	t
T. curucui	Fh	Sc,C	S	F	tt
MOMOTIDAE (1)					
Momotus aequatorialis	Fh	Sc,U	S	U	t
BUCCONIDAE (1)					
Nystalus striolatus	Fh	Sc,C	S	U	t
CAPITONIDAE (1)					
Eubucco versicolor	Fh	Sc,C	S,M	F	t
RAMPHASTIDAE (2)					
Aulacorhynchus derbianus	Fh	C	S	U	t
Ramphastos culminatus	Fh	C	S,G	C	t
PICIDAE (5)					
Picumnus dorbygnianus	Fh,Sg	Sc,C	S,M	U	t
Melanerpes cruentatus	Fe,Sg	C	S	U	si
Veniliornis affinis	Fh	C	M	F	t
Piculus rubiginosus	Fh	C	S,M	F	t
Campephilus rubricollis	Fh	U,Sc	S	F	t
DENDROCOLAPTIDAE (4)					
Sittasomus griseicapillus	Fh	U,Sc	M	F	t
Dendrocolaptes picumnus	Fh	Sc	S,M	U	t
Xiphorhynchus triangularis	Fh	Sc	S,M	U	t
Lepidocolaptes affinis	Fh	C	M	F	t
FURNARIIDAE (10)					
Synallaxis azarae	Fe,Sg	U	S	C	t
Cranioleuca curtata	Fh	C	M	F	t
Premnoplex brunnescens	Fh	U	S,M	F	si

Habitats	
Fh	Montane forest
Fe	Forest edge
B	Bamboo
Sg	Second growth
Foraging Position	
T	Terrestrial
U	Undergrowth
Sc	Subcanopy
C	Canopy
W	Water
A	AeriaL
Sociality	
S	Solitary or in pairs
G	Gregarious
M	Mixed-species flocks
A	Army ant followers
Abundance	
C	Common
F	Fairly common
U	Uncommon
R	Rare
(M)	Migrant
(Mn)	Migrant from north
(Ms)	Migrant from south
Evidence	
sp	Specimen
t	Tape
ph	Photo
si	ID by sight or sound
*	First record
+	First record for Bolivia

APPENDIX 4

	Habitats	Foraging	Sociality	Abundance	Evidence
Hyloctistes subulatus	Fh	Sc	S,M	U	t
Syndactyla rufosuperciliata	Fh,B	U	S,M	U	t
Anabacerthia striaticollis	Fh	Sc,C	M	F	t
Philydor rufus	Fh	C	M	U	si
Thripadectes ignobilis	Fh	U	S,M	U	si
Xenops rutilans	Fh	C	M	F	t
Sclerurus sp.	Fh	T	S	R	t
FORMICARIIDAE (6)					
Thamnophilus palliatus	Sg	U	S	U	t
T. aroyae	Fe,B	U,Sc	S	F	t
Myrmotherula longicauda	Fe	U,Sc	S,M	U	t
Pyriglena leuconota	Fe,Sg	U	S	F	t
Chamaeza campanisona	Fh	T	S	F	t
Conopophaga ardesiaca	Fh	U,T	S	U	t
RHINOCRYPTIDAE (2)					
Scytalopus sp. 1	Fh	T,U	S	C	t
Scytalopus sp. 2	Fh	T,U	S	U	t
COTINGIDAE (4)					
Ampelion rufaxilla	Fh	C	S,M	R	si
Lipaugus vociferans	Fh	Sc,C	S	C	t
Rupicola peruviana	Fh	Sc	S	F	t
Oxyruncus cristatus	Fh	C	S,M	U	si
PIPRIDAE (3)					
Schiffornis turdinus	Fh	U	S	F	t
Piprites chloris	Fh	Sc,C	M	F	t
Chiroxiphia boliviana	Fh,Fe	U	S	C	t
TYRANNIDAE (32)					
Acrocordophus burmeisteri	Fh	C	M	U	t
Zimmerius bolivianus	Fh	C	S,M	C	t
Elaenia flavogaster	Sg	U,C	S	F	t
E. albiceps	Fe,Sg	C	M	F(Ms?)	si
E. obscura	Fh,Fe	C	S,M	C	t
E. pallatangae	Fh	C	S	R	si
Mecocerculus leucophrys	Fh	C	M	U	t
M. hellmayri	Fh	C	M	F	t
Mionectes striaticollis	Fh	Sc,C	S,M	F	si
Leptopogon superciliaris	Fh	Sc,C	M	C	t
Pogonotriccus ophthalmicus	Fh	Sc,C	M	F	t
Phylloscartes ventralis	Fh,Sg	C	S,M	F	t
Paeudotriccus simplex	Fh	U	S	U	si

APPENDIX 4

	Habitats	Foraging	Sociality	Abundance	Evidence
Hemitriccus spodiops	Fe,B	U,Sc	S	F	t
Todirostrum plumbeiceps	Fe,B	U	S	U	t
T. cinereum	Sg	U,C	S	U	si
Platyrinchus mystaceus	Fh	U	S	U	t
Myiophobus fasciatus	Fe,Sg,B	U	S	F	t
Pyrrhomyias cinnamomea	Fh,Fe	A(U/C)	S,M	F	t
Mitrephanes olivaceus	Fh,Fe	A(U/C)	M	F	t
Contopus fumigatus	Fh	A(C)	S,M	U	si
Lathrotriccus euleri	Fh,B	U,Sc	S	U	t
Myiotheretes striaticollis	Sg	A(C)	S	R	t
Myiarchus tuberculifer	Fe	Sc,C	S,M	U	t
M. cephalotes	Fe	Sc,C	S,M	F	t
M. ferox	Sg	Sc,C	S	F	t
Myiozetetes similis	Sg	Sc,C	S	C	t
Myiodynastes chrysocephalus	Fh	C	S,M	U	t
Tyrannus melancholicus	Fe,Sg	A(C)	S	F	si
Pachyramphus polychopterus	Fh,Sg	C	M	U	t
P. castaneus	Fh,Sg	C	S,M	F	t
Tityra semifasciata	Fh	C	S,M	U	t
HIRUNDINIDAE (1)					
Notiochelidon cyanoleuca	Sg	A	G	C	si
CORVIDAE (1)					
Cyanocorax yncas	Fh	Sc,C	G,M	U	t
TROGLODYTIDAE (4)					
Thryothorus genibarbis	Fh,Sg,B	U	S	C	t
Troglodytes aedon	Sg	U	S	C	t
Henicorhina leucophrys	Fh	U	S	C	t
Cyphorhinus thoracicus	Fh	U	S	U	si
TURDIDAE (2)					
Turdus serranus	Fh,Fe	Sc,C	S	F	t
T. leucomelas	Fe,Sg	T,C	S	F	t
VIREONIDAE (2)					
Cyclarhis gujanensis	Fh,Sg	C	S,M	F	t
Smaragdolanius leucotis	Fh	C	M	F	t
ENBERIZINAE (4)					
Zonotrichia capensis	Sg	T,U	S	C	t
Haplospiza rustica	Fe,B	U,Sc	S	U	t
Sporophila caerulescens	Sg	T,U	G	si	
Atlapetes torquatus	Fh	T,U	S	F	t
CARDINALINAE (1)					

Habitats	
Fh	Montane forest
Fe	Forest edge
B	Bamboo
Sg	Second growth
Foraging Position	
T	Terrestrial
U	Undergrowth
Sc	Subcanopy
C	Canopy
W	Water
A	AeriaL
Sociality	
S	Solitary or in pairs
G	Gregarious
M	Mixed-species flocks
A	Army ant followers
Abundance	
C	Common
F	Fairly common
U	Uncommon
R	Rare
(M)	Migrant
(Mn)	Migrant from north
(Ms)	Migrant from south
Evidence	
sp	Specimen
t	Tape
ph	Photo
si	ID by sight or sound
*	First record
+	First record for Bolivia

	Habitats	Foraging	Sociality	Abundance	Evidence
Saltator maximus	Fh,Sg	Sc,C	S,M	C	t
THRAUPINAE (37)					
Schistochlamys melanopis	Fh,Sg	Sc,C	G,M	C	t
Chlorospingus ophthalmicus	Fh	U,C	G,M	C	t
Hemispingus melanotis	Fh,B	U	M	F	t
Thlypopsis ruficeps	Fh,Fe	U,Sc	M	F(M?)	si
Hemithraupis guira	Fh	C	M	F	t
Tachyphonus rufiventer	Fh	C	M	F	t?
Trichothraupis melanops	Fh	Sc,U	S,M	U/R	si
Habia rubica	Fh	U	M	R	si
Piranga flava	Fh	C	S,M	F	t
P. leucoptera	Fh	C	M	F	t
Ramphocelus carbo	Fh,Sg	U/C	G,N	C	t
Thraupis sayaca	Fe,Sg	C	S,M	F	t
T. palmarum	Fh	C	S,M	C	t?
Anisognathus flavinuchus	Fh	C,Sc	M	F	t
Pipraeidea melanonota	Fh,Fe	C,Sc	M	U	si
Euphonia mesochrysa	Fh	C	M	C	t
E. xanthogaster	Fh,Fe	C,U	M	C	t
Chlorophonia cyanea	Fh	C	S,M	F	t
Chlorochrysa calliparaea	Fh	C	M	U	si
Tangara chilensis	Fh,Sg	C	G,M	C	t
T. arthus	Fh	C	M	F	t
T. xanthocephala	Fh	C	M	R	si
T. xanthogastra	Fh	C	M	R	si
T. punctata	Fh	C	M	F	t
T. gyrola	Fh	C	M	R	si
T. meyerdeschauenseei?	Sg	C	S,M	U	si
T. cyanotis	Fh	C	M	C	t
T. cyanicollis	Fh	C	M	C	t
T. nigroviridis	Fh	C	M	R	si
Dacnis lineata	Fh	C	M	F	t
D. cayana	Fh,Sg	C	M	C	si
Chlorophanes spiza	Fh	C	S,M	U	si
Cyanerpes caeruleus	Fh	C	S,M	F	si
Diglossa baritula	Fh,Sg	C,U	S	U	si
D. glauca	Fh	C	M	F	si
Tersina viridis	Fe,Sg	C	S,G	C	t?
Coereba flaveola	Fe,Sg	C	S	U	t

APPENDIX 4

	Habitats	Foraging	Sociality	Abundance	Evidence
PARULINAE (5)					
Parula pitiayumi	Fh	C	M	C	t
Myioborus miniatus	Fh	C,U	M	C	t
Basileuterus bivittata	Fe,B	U	S,M	F	t
B. coronatus	Fh	U	S,M	R	hd
B. tristriatus	Fh	U,Sc	M	C	t
ICTERIDAE (1)					
Psarocolius decumanus	Fh,Sg	C	S,G,M	F	t
CARDUELIDAE (1)					
Carduelis olivacea	Fh	C	S,G	U	si

Habitats	
Fh	Montane forest
Fe	Forest edge
B	Bamboo
Sg	Second growth
Foraging Position	
T	Terrestrial
U	Undergrowth
Sc	Subcanopy
C	Canopy
W	Water
A	AeriaL
Sociality	
S	Solitary or in pairs
G	Gregarious
M	Mixed-species flocks
A	Army ant followers
Abundance	
C	Common
F	Fairly common
U	Uncommon
R	Rare
(M)	Migrant
(Mn)	Migrant from north
(Ms)	Migrant from south
Evidence	
sp	Specimen
t	Tape
ph	Photo
si	ID by sight or sound
*	First record
+	First record for Bolivia

APPENDIX 5

Mammal List

Louise H. Emmons, 1990

	Alto Río Madidi	13 km W. Ixiamas	Calabatea	Rio Machariapo
Opossums				
Caluromys lanatus	X			
Didelphis marsupialis	X			X
Glironia venusta				X
Marmosa murina	X*			
Marmosops noctivagus	X*	X*		
Micoureus cinereus	X*	X*		X*
Metachirus nudicaudatus	X			
Anteaters				
Tamandua tetradactyla			X	
Bats				
Artibeus jamaicensis	X*			
A. literatus	X*			
A. obscura	X*			
Carollia brevicauda	X*			
C. castanea	X*			
C. perspicillata	X*			
Mesophylla macconnelli	X*			
Micronycteris nicefori	X*			
Phyllostomus elongatus	X*			
P. hastatus	X*			
Sturnira lilium	X*			
Primates				
Aotus trivirgatus	X			
Alouatta seniculus	X			
Ateles paniscus	X	X	X	
Callicebus moloch	X			
Cebus apella	X	X		
Saimiri sciureus	X			
Saguinus fuscicollis	X	X		
Carnivores				
Atelocynus microtis	X			
Bassaricyon gabbii	X			
Felis pardalis	X			
Lutra longicaudis	X			
Nasua nasua	X			
Panthera onca	X			
Potos flavus	X	X		
Tapir				
Tapirus terrestris	X			

Specimen collected

	Alto Río Madidi	13 km W. Ixiamas	Calabatea	Río Machariapo
Artiodactyla				
Mazama americana	X			
Tayassu tajacu	X			
Rodents				
Akodon aerosus			X*	X*
Agouti paca	X			
Dasyprocta variegata	X*	X		X
Dactylomys dactylinus	X			
Hydrochaeris hydrochaeris	X			
Mesomys hispidus	X*			
Oecomys bicolor	X*			
Oligoryzomys microtis	X*			
Oryzomys capito		X*		
O. nitidus				X*
Proechimys simonsi	X*			
P. steerei	X*			
Sciurus ignitus	X*			
S. spadiceus	X*			
Rabbit				
Sylvilagus brasiliensis	X			

APPENDIX 6

Observations on the Herpetofauna

Louise H. Emmons, 1990

The small collection of herps identified so far includes no species new to Bolivia. One *Phyllomedusa* collected at both Alto Madidi and Ixiamas is a new species currently being described by Ron Crombie of the Smithsonian Institution. A number of specimens of this species are already known from Bolivia. A *Bufo* of the *typhonius* group awaits revision of the taxon, which is being split.

At Alto Madidi both river turtles (*Podocnemis unifilis*) and white caiman (*Caiman crocodilus*) were common right by the camp, again attesting to the lack of intensive subsistence hunting at this site. *Paleosuchus trigonatus* were common throughout the quebradas in the forest, and even in the roadside ditches, but no *P. palpebrosus* were recognized.

	Alto Río Madidi	13 km W. Ixiamas	Calabatea	Río Machariapo
Frogs				
Bufo poeppigi	X*			
Bufo (*typhonius* group)	X*			
Centrolenella bergeri		X*		
Eleutherodactylus fenestratus	X*			
Eleutherodactylus cf. *discoidales*			X*	
Hyla fasciata	X*			
Hyla geographica	X*			
Hyla lanciformis	X*			
Phyllomedusa sp. nov.	X*	X*		
Phyllomedusa vaillanti	X*			
Rana palmipes	X*			
Snakes				
Bothrops atrox	X*			
Corallus enhydris	X			
Dipsas catesbyi	X*			
Lachesis muta	X			
Lizards				
Ameiva ameiva	X			X
Proctoporus guentheri			X*	
Crocodilians				
Caiman crocodilus	X			
Paleosuchus trigonatus	X			
Turtles				
Podocnemis unifilis	X			

* Specimen collected.

Plant List: Alto Madidi, Bajo Tuichi, and the Foothill Ridges

Robin B. Foster, Alwyn H. Gentry, Stephan Beck, 1990

T	tree (dbh 10 cm, height 5 m)
S	shrub
L	liana
V	herbaceous vine
H	herb
E	epiphyte

The list compiled here is a combination of the field lists of plants observed by R. Foster with the plant collection lists of A. Gentry and S. Beck. These identifications are based on the experience of the authors and made without direct benefit of herbarium comparisons, published references, or detailed study. Most were neither flowering nor fruiting. They are certainly at least 90-95% correct, but should still be used with caution. Many species occur in a wide range of habitats but are listed here only once in the habitat of their apparent greatest abundance or greatest relative importance.

Sandy-Silty Beach

COMPOSITAE	
Baccharis salicifolia	S
Tessaria integrifolia	S
GRAMINEAE	
Gynerium saggitatum	S
SALICACEAE	
Salix humboldtiana	T

Stable River Bank

AMARANTHACEAE	
Iresine sp.	H
CAPPARIDACEAE	
Cleome sp.	H
Cleome spinosa	H
COMPOSITAE	
Spilanthes sp.	H
ELAEOCARPACEAE	
Muntingia calabura	S
GRAMINEAE	
Imperata sp.	H
LEGUMINOSAE-MIM	
Calliandra angustifolia	S
Pithecellobium longifolium	T
LYTHRACEAE	
Adenaria floribunda	S
MALVACEAE	
? sp.	H

Young Floodplain

ACANTHACEAE	
Sanchezia peruviana	S
? sp.	S
? sp.	S
AMARANTHACEAE	
Chamissoa altissima	H
ANNONACEAE	
Guatteria acutissima cf.	T
BOMBACACEAE	
Ochroma pyramidale	T
BORAGINACEAE	
Cordia alliodora	T
CHRYSOBALANACEAE	
Licania britteniana	T
COMBRETACEAE	
Terminalia oblonga	T
EUPHORBIACEAE	
Acalypha macrostachya	S
Acalypha mapirensis	S
Margaritaria nobilis	*T*
Sapium aereum cf.	*T*
Sapium ixiamasense	T
Sapium marmieri	T
LAURACEAE	
Nectandra reticulata	T
LEGUMINOSAE-CAES	
Senna sp.	S
LEGUMINOSAE-MIM	
Acacia loretensis	T

Young Floodplain
continued

Inga marginata	T
Inga ruiziana	T
Piptadenia sp.	L
LEGUMINOSAE-PAP	
Erythrina poeppigiana	T
Erythrina ulei	T
MARANTACEAE	
Calathea capitata	H
MELASTOMATACEAE	
Miconia aulocalyx	S
MELIACEAE	
Cedrela odorata	T
Guarea guidonia	T
Trichilia quadrijuga	T
MORACEAE	
Cecropia membranacea	T
Ficus insipida	T
Ficus maxima	T
Sorocea pileata	T
MUSACEAE	
Heliconia episcopalis	H
PIPERACEAE	
Piper longestylosum	S
Piper obovatum	S
POLYGONACEAE	
Triplaris americana	T
RUBIACEAE	
Calycophyllum spruceanum	T
SOLANACEAE	
Cestrum reflexum cf.	S
Cuatresia fosteri	S
Solanum robustifrons	S
Solanum viridipes	S
STERCULIACEAE	
Byttneria aculeata	L
Byttneria pescapraeifolia	L
Byttneria sp.	S
Guazuma crinita	T
TILIACEAE	
Heliocarpus americana	T
URTICACEAE	
Boehmeria sp.	S
Urera baccifera	L
Urera laciniata	S
VERBENACEAE	
Citharexylum poeppigii	T
VITACEAE	
Cissus sicyoides	L
ZINGIBERACEAE	
Costus scaber	H
Renealmia thyrsoidea	H

Older Floodplain

ACANTHACEAE	
Aphelandra aurantiaca	S
Justicia sp.	H
Pachystachys spicata	S
Ruellia thyrsostachya	S
ANACARDIACEAE	
Spondias mombin	T
ANNONACEAE	
Crematosperma sp.	T
Oxandra acuminata	T
Rollinia pittieri	T
Ruizodendron ovale	T
Unonopsis floribunda	T
Xylopia cuspidata	S
APOCYNACEAE	
Aspidosperma sp.	T
Pacouria boliviensis	L
Stenosolen sp.	L
Tabernaemontana sp.	T
ARACEAE	
Dieffenbachia humilis	H
Dracontium loretense	H

Species	Habit
Monstera obliqua	E
Philodendron sp.	E
Philodendron acreanum	E
Syngonium sp.	E
ARALIACEAE	
Dendropanax arboreus	T
BIGNONIACEAE	
Arrabidaea patellifera	L
Arrabidaea verrucosa	L
Mansoa standleyi	L
Tynnanthus schumannianus	L
BIXACEAE	
Bixa urucurana	T
BOMBACACEAE	
Ceiba pentandra	T
Quararibea cordata	T
Quararibea rhombifolia	T
Quararibea wittii	T
BORAGINACEAE	
Cordia sp.	T
Cordia nodosa	S
BURSERACEAE	
Tetragastris sp.	T
CAMPANULACEAE	
Centropogon cornutus	H
CAPPARIDACEAE	
Morisonia oblongifolia	S
CARICACEAE	
Carica microcarpa	S
Jacaratia digitata	T
CELASTRACEAE	
Maytenus magnifolia	T
CHRYSOBALANACEAE	
Hirtella sp.	T
Licania heteromorpha	T
Parinari parilis cf.	T
COMBRETACEAE	
Combretum assimile	L
Combretum laxum	L
COMMELINACEAE	
Campelia zanonia	H
Dichorisandra sp.	H
CUCURBITACEAE	
Fevillea cordifolia	L
CYPERACEAE	
Rhynchospora umbraticola	H
DILLENIACEAE	
Tetracera parviflora	L
ELAEOCARPACEAE	
Sloanea guianensis	T
Sloanea obtusifolia cf.	T
EUPHORBIACEAE	
Acalypha diversifolia	S
Apodandra brachybotrya	V
Croton tessmannii	T
Drypetes sp.	S
Hura crepitans	T
Mabea maynensis	T
FLACOURTIACEAE	
Casearia combaymensis cf.	S
Hasseltia floribunda	T
Lacistema sp.	S
Lunania parviflora	T
Mayna odorata	S
GESNERIACEAE	
Codonanthe uleana	E
Gloxinia sp.	H
GRAMINEAE	
Pharus sp.	H
GUTTIFERAE	
Chrysochlamys sp.	T
HIPPOCRATEACEAE	
Cheiloclinium sp.	T
Peritassa huanucana cf.	L
Salacia sp.	L
Salacia macrantha	T
ICACINACEAE	
Calatola venezuelana	T

T	tree (dbh 10 cm, height 5 m)
S	shrub
L	liana
V	herbaceous vine
H	herb
E	epiphyte

Older Floodplain
continued

LAURACEAE	
Endlicheria dysodantha	T
LEGUMINOSAE-CAES	
Bauhinia glabra	L
LEGUMINOSAE-MIM	
Inga sapindoides	T
Pithecellobium latifolium	T
LEGUMINOSAE-PAP	
Andira inermis	T
Dipteryx micrantha	T
Lecointea peruviana	T
LORANTHACEAE	
Oryctanthus sp.	E
MALPIGHIACEAE	
Tetrapteris sp.	L
MARANTACEAE	
Calathea crotalifera	H
Calathea lutea	H
Calathea micans	H
Calathea sp.	H
Monotagma sp.	H
MELASTOMATACEAE	
Miconia triplinervis	S
MELIACEAE	
Trichilia maynasiana	T
Trichilia pleeana	T
Trichilia sp.	T
MENISPERMACEAE	
Anomospermum grandifolium	L
MONIMIACEAE	
Mollinedia racemosa	S
Siparuna sp.	S
MORACEAE	
Clarisia biflora	T
Clarisia racemosa	T
Ficus sp.	T
Ficus killipii	T
Ficus paraensis	T
Ficus perforata	T
Naucleopsis krukovii cf.	T
Poulsenia armata	T
Pseudolmedia laevis	T
Sorocea steinbachii cf.	T
MUSACEAE	
Heliconia metallica	H
MYRISTICACEAE	
Otoba parvifolia	T
Virola sebifera	T
MYRSINACEAE	
Cybianthus sp.	S
MYRTACEAE	
Psidium friedrichsthalianum cf.	T
? sp.	S
NYCTAGINACEAE	
Pisonia aculeata	L
OLACACEAE	
Heisteria acuminata	T
ORCHIDACEAE	
? sp.	H
PALMAE	
Astrocaryum macrocalyx	T
Bactris actinoneura	S
Chamaedorea sp.	S
Euterpe precatoria	T
Geonoma sp.	S
Iriartea deltoidea	T
Socratea exorrhiza	T
PHYTOLACCACEAE	
Petiveria alliacea	S
Trichostigma octandra	L
PIPERACEAE	
Peperomia macrostachya	E
Piper laevigatum	S
POLYGONACEAE	
Coccoloba sp.	S
Triplaris poeppigiana	T

PTERIDOPHYTA	
Lomariopsis japurensis	E
Polypodium polypodioides	E
RHAMNACEAE	
Gouania sp.	L
RUBIACEAE	
Genipa americana	T
Geophila macropoda	H
Hamelia axillaris	S
Ixora peruviana	S
Macrocnemum roseum	T
Psychotria carthaginensis	S
Randia armata	S
Randia ruiziana	S
Randia sp. nov.	S
SAPINDACEAE	
Allophylus glabratus	S
Sapindus saponaria	T
SAPOTACEAE	
Pouteria pariry cf.	T
Sarcaulus brasiliensis	T
SIMAROUBACEAE	
Picramnia latifolia	S
SMILACACEAE	
Smilax febrifuga	L
STAPHYLEACEAE	
Huertea glandulosa	T
Turpinia occidentalis	T
STERCULIACEAE	
Herrania sp.	S
THEOPHRASTACEAE	
Clavija reflexiflora	S
TILIACEAE	
Luehea cymulosa	T
ULMACEAE	
Celtis iguanea	L
Celtis schippii	T
URTICACEAE	
Pouzolzia sp.	S
VERBENACEAE	
Aegiphila cuneata	S
Aegiphila haughtii	S
VIOLACEAE	
Leonia sp.	T
ZINGIBERACEAE	
Renealmia sp.	H

High Terrace & Slopes

ACANTHACEAE	
Aphelandra sp.	S
Aphelandra goodspeedii	H
Mendoncia sp.	V
Mendoncia sp.	V
Ruellia graecizans	S
Ruellia tarapotana	S
? sp.	S
ANACARDIACEAE	
Astronium graveolens	T
ANNONACEAE	
Anaxagorea sp.	T
Crematosperma leiophylla	S
Duguetia quitarensis	T
Oxandra sp.	T
Oxandra espintana	S
Rollinia sp.	T
Xylopia sp.	T
APOCYNACEAE	
Aspidosperma marcgravianum	T
Aspidosperma tambopatense	T
Aspidosperma vargasii	T
Forsteronia sp.	L
Forsteronia sp.	L
Himatanthus sucuuba	T
Tabernaemontana sp.	S
Tabernaemontana sp.	T
ARACEAE	
Anthurium sp.	H

T	tree (dbh 10 cm, height 5 m)
S	shrub
L	liana
V	herbaceous vine
H	herb
E	epiphyte

High Terrace & Slopes
continued

Anthurium clavigerum	E
Anthurium croatii	E
Anthurium eminens cf.	E
Anthurium ernestii	E
Anthurium kunthii cf.	E
Dieffenbachia sp.	H
Heteropsis oblongifolia	V
Monstera dubia	E
Monstera sp.	E
Monstera subpinnata	E
Philodendron sp.	E
Philodendron sp.	V
Philodendron acreanum	E
Philodendron ernestii	E
Rhodospatha sp.	E
Xanthosoma sp.	H
Xanthosoma pubescens	H
ARISTOLOCHIACEAE	
Aristolochia sp.	L
ASCLEPIADACEAE	
? sp..	L
? sp..	L
BEGONIACEAE	
Begonia sp..	H
Begonia glabra	V
BIGNONIACEAE	
Adenocalymma impressum	L
Adenocalymma purpurascens	L
Adenocalymma uleanum	L
Anemopaegma chrysoleucum	L
Arrabidaea affinis	L
Arrabidaea chica	L
Arrabidaea pearcei	L
Arrabidaea platyphylla	L
Ceratophytum tetragonolobum	L
Clytostoma sciuripabulum	L
Cuspidaria floribunda	L
Cuspidaria lateriflora	L
Cydista aequinoctialis	L
Cydista lilacina	L
Distictis occidentalis	L
Jacaranda copaia	T
Jacaranda glabra	T
Macfadyena uncata	L
Mansoa parvifolia	L
Mansoa verrucifera	L
Mussatia hyacintha	L
Roentgenia bracteata	L
Roentgenia bracteomana	L
Spathicalyx xanthophylla	L
Stizophyllum riparium	L
Tabebuia incana	T
Tanaecium nocturnum	L
Tynnanthus sp.	L
Xylophragma pratense	L
BOMBACACEAE	
Cavanillesia umbellata	T
Chorisia sp.	T
Eriotheca globosa	T
Huberodendron swietenioides	T
Pachira sp.	T
Pseudobombax septenatum	T
BORAGINACEAE	
Cordia sp.	T
BROMELIACEAE	
Bromelia? sp.	H
Streptocalyx longifolia	E
BURSERACEAE	
Protium sp. 1	T
Protium sp. 2	T
Protium sp. 3	T
Protium unifoliolatum	T
Tetragastris altissima	T
Trattinnickia	T
CACTACEAE	
Epiphyllum phyllanthus cf.	E
CAMPANULACEAE	
Centropogon sp.	H

CAPPARIDACEAE	
Capparis sola cf.	S
Cleome sp.	S
CARYOCARACEAE	
Anthodiscus sp.	T
Caryocar amygdaliforme	T
CHRYSOBALANACEAE	
Hirtella bullata cf.	T
Hirtella racemosa	S
Hirtella sp..	T
Licania hypoleuca	T
Licania sp.	T
Licania sp.	T
Parinari klugii	T
COMBRETACEAE	
Buchenavia sp.	T
Combretum sp.	L
Combretum sp.	L
Combretum sp.	L
Terminalia amazonica	T
COMMELINACEAE	
Floscopa elegans	H
COMPOSITAE	
Vernonia brachiata	S
CONNARACEAE	
Connarus punctatus cf.	L
? sp.	L
CONVOLVULACEAE	
Maripa peruviana	L
CUCURBITACEAE	
Cayaponia sp.	L
Gurania spinulosa	L
? sp.	L
? sp.	L
CYCLANTHACEAE	
Asplundia sp.	E
Carludovica almata	H
Cyclanthus bipartitus	H
Thoracocarpus bissectus	E

CYPERACEAE	
Diplasia karatifolia	H
Scleria secans	V
DILLENIACEAE	
Doliocarpus sp.	L
Tetracera sp.	L
? sp.	L
DIOSCOREACEAE	
Dioscorea acanthogyne	L
Dioscorea sp.	L
ELAEOCARPACEAE	
Sloanea fragrans	T
Sloanea sp.	T
Sloanea sp.	T
EUPHORBIACEAE	
Acalypha obovata	T
Acalypha sp.	L
Alchornea glandulosa	T
Croton matourensis	T
Drypetes sp.	T
Drypetes amazonica	T
Hevea brasiliensis	T
Hyeronima alchorneoides	T
Mabea sp.	T
Manihot sp.	V
Omphalea diandra	L
Pausandra trianae	S
Richeria racemosa	T
Senefeldera sp.	T
Tetrorchidium sp.	T
? sp.	T
FLACOURTIACEAE	
Banara guianensis	T
Casearia javitensis	T
Casearia obovalis	S
Lacistema sp.	S
Lacistema aggregatum	S
Laetia procera	T
Lindackeria paludosa	T
Tetrathylacium macrophyllum	T

T	tree (dbh 10 cm, height 5 m)
S	shrub
L	liana
V	herbaceous vine
H	herb
E	epiphyte

High Terrace & Slopes
continued

GENTIANACEAE	
Voyria sp.	H
GESNERIACEAE	
Besleria sp.	H
Drymonia semicordata	V
GRAMINEAE	
Arundinella berteroniana	H
Lasiacis sp.	H
Olyra sp.	H
Pariana sp.	H
GUTTIFERAE	
Calophyllum brasiliense	T
Chrysochlamys ulei cf.	T
Garcinia (Rheedia) madruno	T
Garcinia (Rheedia) brasiliensis	T
Marila sp.	T
Quapoya peruviana	L
Symphonia globulifera	T
Tovomita stylosa	S
Vismia sp.	S
HAEMODORACEAE	
Xiphidium caeruleum	H
HIPPOCRATEACEAE	
Anthodon decussatum	L
Salacia sp.	T
Salacia sp.	L
ICACINACEAE	
Calatola sp.	T
Citronella incarum	T
Discophora guianensis	T
LAURACEAE	
Aniba sp.	T
Caryodaphnopsis fosteri	T
Endlicheria sp.	T
Licaria triandra cf.	T
Ocotea cernua	T
Persea sp.	T
Pleurothyrium krukovii	T
? sp.	T
? sp.	T
? sp.	T
? sp.	T
? sp.	T
? sp.	T
? sp.	T
? sp.	T
? sp.	T
? sp.	T
? sp.	T
? sp.	T
? sp.	T
LECYTHIDACEAE	
Cariniana decandra	T
Couratari guianensis	T
Eschweilera sp.	T
Eschweilera sp.	T
LEGUMINOSAE-CAES	
Bauhinia guianensis	L
Bauhinia hirsutissima	L
Copaifera reticulata	T
Hymenaea oblongifolia	T
Swartzia sp.	T
Swartzia sp.	T
Tachigali sp.	T
Tachigali sp.	T
LEGUMINOSAE-MIM	
Acacia sp.	T
Cedrelinga catenaeformis	T
Inga stipularis	T
Inga sp. a	T
Inga sp. b	T
Inga sp. c	T
Inga sp. d	T
Inga sp. e	T
Inga sp. f	T
Inga sp. g	T
Inga sp. h	T

Species	Habit
Parkia sp.	T
Pithecellobium sp.	T
Pithecellobium sp.	T
Pithecellobium macrophyllum	T
LEGUMINOSAE-PAP	
Amburana cearensis	T
Apuleia leiocarpa	T
Canavalia sp.	L
Clitoria sp.	L
Dalbergia loretana	T
Dialium guianense	T
Diplotropis? sp.	T
Dussia tessmannii cf.	T
Machaerium cuspidatum	L
Machaerium kegelii	T
Myroxylon balsamum	T
Ormosia sp.	T
LOGANIACEAE	
Potalia amara	S
Strychnos sp. a	L
Strychnos sp. b	L
Strychnos sp. c	L
Strychnos tarapotensis	S
LORANTHACEAE	
Struthanthus sp.	E
LYTHRACEAE	
Lafoensia sp.	T
MALPIGHIACEAE	
Hiraea sp.	L
? sp.	T
? sp.	T
MARANTACEAE	
Calathea roseopicta cf.	H
Hylaeanthe unilateralis	H
Ischnosiphon sp.	H
Ischnosiphon puberulus	V
MARCGRAVIACEAE	
Marcgravia sp.	L
MELASTOMATACEAE	
Blakea sp.	T
Clidemia heterophylla	S
Clidemia septuplinervia	S
Leandra longicoma	S
Miconia sp.	S
Miconia sp.	T
Miconia sp.	T
Miconia sp.	T
Miconia sp.	T
Miconia affinis cf.	T
Miconia bubalina	S
Miconia nervosa	S
Miconia paleacea	S
Miconia procumbens	S
Miconia sp.	T
Miconia sp.	S
Miconia tomentosa cf.	T
Miconia triplinervis	S
Mouriri myrtilloides	S
Mouriri peruviana cf.	S
Tococa guianensis	S
Tococa parviflora	S
MELIACEAE	
Cabralea cangerana	T
Guarea sp.	T
Guarea kunthiana	S
Guarea pterorachis	T
Ruagea sp.	T
Ruptiliocarpon sp. nov.	T
Trichilia elegans	T
Trichilia pachypoda cf.	T
Trichilia pallida	T
Trichilia septentrionalis	T
Trichilia solitudinus	T
MENISPERMACEAE	
Anomospermum sp.	L
Chondodendron sp.	L
Odontocarya sp.	L
? sp.	L
MONIMIACEAE	

T	tree (dbh 10 cm, height 5 m)
S	shrub
L	liana
V	herbaceous vine
H	herb
E	epiphyte

High Terrace & Slopes
continued

Mollinedia sp.	S
Mollinedia sp.	S
Siparuna sp.	S
Siparuna decipiens	T
Siparuna sp.	S
Siparuna sp.	S
MORACEAE	
Batocarpus amazonicus	T
Brosimum alicastrum	T
Brosimum lactescens cf.	T
Brosimum sp.	T
Brosimum sp.	T
Castilla ulei	T
Cecropia sp.	T
Cecropia polystachya	T
Cecropia sciadophylla	T
Coussapoa sp.	E
Coussapoa ovalifolia	E
Dorstenia umbricola cf.	S
Ficus sp.	T
Ficus sp.	T
Ficus citrifolia	T
Ficus gommeleira	T
Ficus juruensis	E
Ficus mathewsii	T
Ficus popenoei	T
Ficus schultesii	T
Ficus sphenophylla	T
Ficus tonduzii	T
Helicostylis tomentosa	T
Maquira calophylla	T
Naucleopsis sp.	T
Naucleopsis terstroemiiflora	T
Olmedia aspera	T
Perebea humilis	S
Perebea sp.	T
Perebea ? sp.	T
Pourouma cecropiifolia	T
Pourouma guianensis	T
Pourouma minor	T
Pseudolmedia laevigata	T
Sorocea guilleminiana	T
MUSACEAE	
Heliconia densiflora	H
Heliconia pruinosa	H
Heliconia sp.	H
Heliconia spathocircinnata	H
Heliconia subulata	H
MYRISTICACEAE	
Iryanthera sp.	T
Iryanthera juruensis	T
Virola calophylla	T
Virola flexuosa	T
Virola mollissima	T
? sp.	T
MYRSINACEAE	
Ardisia guianensis cf.	S
Cybianthus sp.	S
Myrsine pellucida	T
Parathesis sp.	S
Stylogyne cauliflora	S
MYRTACEAE	
Calyptranthes densiflora	S
Calyptranthes lanceolata aff.	S
Campomanesia sp.	S
Eugenia sp.	T
Eugenia sp.	S
Eugenia sp.	S
Eugenia sp.	S
Myrcia sp.	T
? sp.	S
? sp.	T
? sp.	T
? sp.	S
NYCTAGINACEAE	
Guapira sp.	T
Neea sp.	T
Neea boliviana	T
Neea spruceana	S

Species	Habit
Neea verticillata	T
OCHNACEAE	
Ouratea iquitosensis cf.	S
OLACACEAE	
Heisteria scandens	L
Minquartia guianensis	T
ORCHIDACEAE	
Huntleya sp.	E
OXALIDACEAE	
Biophytum sp.	H
PALMAE	
Aiphanes sp.	S
Bactris sp.	S
Chamaedorea pinnatifrons	S
Geonoma sp.	S
Geonoma sp.	S
Geonoma deversa	S
Geonoma sp.	S
Hyospathe elegans	S
Jessenia batahua	T
Oenocarpus mapora	T
Phytelephas macrocarpa	S
Wendlandiella sp.	S
Wettinia sp.	T
PASSIFLORACEAE	
Passiflora auriculata	L
Passiflora coccinea	L
PIPERACEAE	
Peperomia serpens	E
Piper augustum	S
Piper costatum	S
Piper crassinervium	S
Piper longifolium	S
Piper obliqum cf.	S
Piper sp.	S
Piper sp.	S
POLYGALACEAE	
Moutabea aculeata	L
Polygala gigantea	H
Securidaca sp.	L
POLYGONACEAE	
Coccoloba sp.	L
PROTEACEAE	
Panopsis? sp.	T
Roupala montana	T
PTERIDOPHYTA	
Adiantum sp.	H
Asplenium serratum	H
Campyloneurum sp.	E
Cyclopeltis semicordata	H
Danaea nodosa	H
Didymochlaena truncatula	H
Hemidictyum marginatum	H
Metaxya rostrata	H
Olfersia cervina	E
Phlebodium decumanum	E
Polybotrya sp.	E
Polypodium sp.	E
Pteris sp.	H
Selaginella sp.	H
Selaginella sp.	H
? sp.	E
? sp.	H
? sp.	H
'treefern' sp.	S
'treefern' sp.	S
'treefern' sp.	S
QUIINACEAE	
Lacunaria sp.	T
Quiina sp.	S
Quiina peruviana cf.	T
RHAMNACEAE	
Zizyphus cinnamomum	T
RUBIACEAE	
Alibertia tutumilla	T
Alseis reticulata	T
Bathysa sp.	T
Bertiera guianensis	S
Calycophyllum acreanum	T

T	tree (dbh 10 cm, height 5 m)
S	shrub
L	liana
V	herbaceous vine
H	herb
E	epiphyte

High Terrace & Slopes
continued

Capirona decorticans	T
Cephaelis dolichophylla	S
Cephaelis tomentosa	S
Chimarrhis hookeri	T
Chomelia klugii	S
Coussarea sp.	T
Coussarea sp.	T
Coussarea sp.	S
Coussarea sp.	S
Faramea angustifolia	S
Faramea anisocalyx cf.	S
Faramea multiflora	S
Palicourea punicea	S
Palicourea subspicata	S
Pentagonia sp.	T
Posoqueria latifolia	T
Psychotria brachiata	S
Psychotria marginata	S
Psychotria officinalis	S
Psychotria racemosa	S
Psychotria sp.	S
Psychotria sp.	S
Psychotria viridis	S
Rudgea cornifolia	S
Rudgea sp.	S
Uncaria sp.	L
Warscewiczia cordata	T
RUTACEAE	
Esenbeckia sp.	S
Galipea trifoliata	S
Metrodorea flavida cf.	T
Raputia sp.	T
Zanthoxylum acreanum	T
Zanthoxylum weberbaueri aff.	T
SABIACEAE	
Ophiocaryon ? sp.	T
SAPINDACEAE	
Allophylus sp.	T
Cupania sp.	T
Matayba sp.	T
Paullinia bracteosa cf.	L
Paullinia sp.	L
Talisia sp.	T
Talisia princeps cf.	T
SAPOTACEAE	
Manilkara inundata cf.	T
Micropholis guyanensis	T
Micropholis sp.	T
Pouteria sp.	T
Pouteria sp.	T
Pouteria sp.	T
Pouteria torta	T
SIMAROUBACEAE	
Picramnia sp.	S
? sp.	T
SMILACACEAE	
Smilax sp.	L
SOLANACEAE	
Cestrum megalophyllum	S
Cyphomandra sp.	S
Lycianthes glandulosum cf.	V
Lycianthes sp.	S
Solanum sp.	S
STERCULIACEAE	
Guazuma ulmifolia	T
Pterygota amazonica	T
Sterculia sp.	T
Theobroma cacao	T
Theobroma speciosa	T
THEOPHRASTACEAE	
Clavija hookeri cf.	S
Clavija longifolia	S
TILIACEAE	
Apeiba membranacea	T
Apeiba tibourbou	T
TRIGONIACEAE	
Trigonia sp.	L

ULMACEAE	
Ampelocera edentula	T
Trema micrantha	T
VERBENACEAE	
Petrea maynensis	L
Vitex sp.	T
VIOLACEAE	
Gloeospermum sp.	S
Leonia sp.	T
Rinorea guianensis	T
Rinoreocarpus ulei	T
VITACEAE	
Cissus sp.	L
VOCHYSIACEAE	
Qualea sp.	T
Vochysia sp.	T
ZINGIBERACEAE	
Costus acreanus cf.	H
Costus sp.	H
Dimerocostus strobilaceus	H
Renealmia sp.	H

Dry & Wet Ridges

AMARANTHACEAE	
Iresine sp.	L
AMARYLLIDACEAE	
Eucharis sp.	H
ANACARDIACEAE	
Mauria? sp.	T
ANNONACEAE	
Guatteria sp.	T
? sp.	T
ARACEAE	
Philodendron sp.	E
ARALIACEAE	
Didymopanax morototoni	T
Schefflera sp.	E
BIGNONIACEAE	
Arrabidaea corallina	L
Arrabidaea florida	L
Lundia sp.	L
Paragonia pyramidata	L
Pithecoctenium crucigerum	L
Pleonotoma melioides	L
BOMBACACEAE	
Pseudobombax sp.	T
BORAGINACEAE	
Tournefortia sp.	L
CELASTRACEAE	
Maytenus sp.	T
COMBRETACEAE	
Combretum sp.	L
COMPOSITAE	
Vernonia sp.	S
CONNARACEAE	
Rourea sp.	L
CONVOLVULACEAE	
? sp.	L
DICHAPETALACEAE	
Tapura juruana	T
DILLENIACEAE	
Davilla nitida	L
DIOSCOREACEAE	
Dioscorea sp.	L
EBENACEAE	
Diospyros sp.	T
ERYTHROXYLACEAE	
Erythroxylum sp.	S
EUPHORBIACEAE	
Acidoton venezolanus	S
Caryodendron orinocense	T
Croton sp.	T
Drypetes sp.	T
Manihot sp.	S
Sapium sp.	T
FLACOURTIACEAE	

T	tree (dbh 10 cm, height 5 m)
S	shrub
L	liana
V	herbaceous vine
H	herb
E	epiphyte

Dry & Wet Ridges
continued

Hasseltia? sp.	T
Homalium? sp.	T
Pleuranthodendron sp.	T
Prockia crucis	S
GRAMINEAE	
Lasiacis sp.	H
GUTTIFERAE	
Clusia sp.	T
Garcinia (Rheedia) sp.	T
Tovomita sp.	T
HIPPOCRATEACEAE	
? sp.	L
LAURACEAE	
Phoebe sp.	T
? sp.	T
? sp.	T
? sp.	T
? sp.	T
? sp.	T
LEGUMINOSAE-CAES	
Bauhinia tarapotensis	T
Hymenaea courbaril	T
LEGUMINOSAE-MIM	
Adenopodia polystachya	L
Calliandra sp.	T
LEGUMINOSAE-PAP	
Dalbergia monetaria	L
Platymiscium sp.	T
LOGANIACEAE	
Strychnos sp.	L
MALPIGHIACEAE	
? sp.	L
? sp.	L
MARANTACEAE	
Calathea peruviana cf.	H
Monotagma sp.	H
MARCGRAVIACEAE	
Norantea sp.	L
Souroubea sp.	L
MELASTOMATACEAE	
Miconia sp.	T
Mouriri sp.	S
MELIACEAE	
Guarea sp.	T
Guarea sp.	T
Trichilia sp.	T
Trichilia sp.	T
Trichilia sp.	T
MENISPERMACEAE	
Abuta grandifolia	L
Abuta sp.	L
Chondodendron tomentosum	L
? sp.	L
MONIMIACEAE	
Mollinedia sp.	S
Siparuna sp.	S
MORACEAE	
Brosimum guianense	T
Clarisia ilicifolia	T
Pseudolmedia macrophylla	T
Sorocea sp.	S
MUSACEAE	
Heliconia sp.	H
Heliconia sp.	H
MYRISTICACEAE	
Virola loretensis	T
MYRSINACEAE	
Ardisia weberbaueri cf.	S
MYRTACEAE	
Calyptranthes sp.	T
Eugenia sp.	T
? sp.	T
? sp.	T
NYCTAGINACEAE	
Neea sp.	T
OLACACEAE	

Species	Habit
Agonandra sp.	T
Heisteria ovata cf.	T
PALMAE	
Bactris sp.	S
Desmoncus sp.	L
Desmoncus sp.	L
Scheelea sp.	T
PIPERACEAE	
Piper callosum cf.	S
POLYGONACEAE	
Coccoloba mollis	T
Coccoloba sp.	L
PTERIDOPHYTA	
Adiantum sp.	H
Nephrolepis sp.	E
Tectaria incisa	H
? sp.	H
? sp.	H
RHAMNACEAE	
Gouania sp.	L
Rhamnidium elaeocarpum	T
RUBIACEAE	
Alibertia pilosa cf.	S
Amaioua corymbosa	T
Chomelia sp.	
Condaminea corymbosa	T
Faramea quinqueflora cf.	S
Gonzalagunia sp.	S
Palicourea macrobotrys	S
Posoqueria sp.	T
Psychotria sp.	S
Randia sp.	T
Rudgea sp.	T
Rustia rubra	S
Simira sp.	
? sp.	T
RUTACEAE	
Erythrochiton sp.	S
Esenbeckia sp.	S
Galipea jasminiflora	S

Species	Habit
Pilocarpus sp.	S
SAPINDACEAE	
Allophylus divaricatus	T
Paullinia sp.	L
Paullinia sp.	L
Paullinia sp.	L
? sp.	T
SAPOTACEAE	
Chrysophyllum sp.	T
Pouteria sp.	T
Pouteria sp.	T
Pouteria sp.	T
? sp.	T
? sp.	T
SMILACACEAE	
Smilax sp.	L
SOLANACEAE	
Lycianthes amatitlanensis cf.	S
Solanum sp.	S
STERCULIACEAE	
Reevesia smithii	T
STYRACACEAE	
Styrax sp.	T
THEOPHRASTACEAE	
Clavija sp.	S
THYMELEACEAE	
? sp.	T
VERBENACEAE	
Aegiphila cordata	L
Citharexylum sp.	T
Lantana camara	S
VIOLACEAE	
Rinorea lindeniana	S
Rinorea viridifolia	S
VITACEAE	
Cissus sp.	L

T	tree (dbh 10 cm, height 5 m)
S	shrub
L	liana
V	herbaceous vine
H	herb
E	epiphyte

Quartzite Ridges

Family / Species	
ANNONACEAE	
Guatteria sp.	T
APOCYNACEAE	
Aspidosperma sp.	T
Aspidosperma sp.	T
ASCLEPIADACEAE	
? sp.	V
? sp.	V
BIGNONIACEAE	
Amphilophium sp.	L
Distictella elongata	L
Pyrostegia dichotoma	L
COMBRETACEAE	
Buchenavia sp.	T
COMPOSITAE	
? sp.	S
CYPERACEAE	
Scleria sp.	H
DIOSCOREACEAE	
Dioscorea sp.	L
ELAEOCARPACEAE	
Sloanea sp.	T
EUPHORBIACEAE	
Alchornea sp.	T
Aparisthmium cordatum	T
Mabea sp.	T
Mabea sp.	T
Maprounea sp.	T
GENTIANACEAE	
Voyria sp.	H
GRAMINEAE	
Pariana sp.	H
GUTTIFERAE	
Clusia sp.	T
HUMIRIACEAE	
Sacoglottis sp.	T
LAURACEAE	
? sp.	T
MELASTOMATACEAE	
Clidemia sp.	S
Graffenriedia sp.	T
Miconia sp.	T
Miconia sp.	T
MYRTACEAE	
? sp. a	T
? sp. b	T
? sp. c	T
? sp. d	T
? sp. e	T
POLYGALACEAE	
Securidaca sp.	L
POLYGONACEAE	
Bredemeyera sp.	L
PTERIDOPHYTA	
Adiantum sp.	H
Asplenium rutaceum	H
Dicranopteris sp.	V
Lindsaea sp.	H
? sp.	V
RUBIACEAE	
Bathysa sp.	T
Cinchona sp.	T
Geophila repens	H
Psychotria sp.	S
Psychotria deflexa	S
SAPINDACEAE	
? sp.	T
STYRACACEAE	
Styrax guianensis cf.	T
THEACEAE	
Freziera sp.	T
ULMACEAE	
Ampelocera sp.	T

Weeds

Family / Species	Habit
APOCYNACEAE	
Mandevilla hirsuta	V
Mesechites trifida	V
BORAGINACEAE	
Heliotropium indicum	H
COMMELINACEAE	
Commelina sp.	H
COMPOSITAE	
Eclipta alba	H
Pseudoelephantopus spiralis	H
CONVOLVULACEAE	
Ipomoea sp.	V
CYPERACEAE	
Cyperus laxus	H
Cyperus odoratus	H
Fimbristylis dichotoma	H
Rhynchospora sp.	H
Scleria sp.	H
EUPHORBIACEAE	
Chamaesyce sp.	H
GENTIANACEAE	
Irlbachia alata	H
GRAMINEAE	
Andropogon bicornis	H
Axonopus sp.	H
Eriochloa sp.	H
Hymenachne amplexifolia	H
Hymenachne donacifolia	H
Leptochloa sp.	H
Panicum laxum cf.	H
Paspalum sp.	H
Paspalum sp.	H
Setaria sp.	H
HYDROPHYLLACEAE	
Nama sp.	H
LABIATAE	
Hyptis sp.	H
Marsypianthes sp.	H
Salvia sp.	H
LEGUMINOSAE-PAP	
Aeschynomene sp.	H
Chaetocalyx brasiliensis	V
Crotalaria nitens	H
Desmodium sp.	H
Desmodium cajanifolium	S
Stylosanthes sp.	H
LOGANIACEAE	
Mitreola petiolata	H
LYTHRACEAE	
Cuphea sp.	H
MALVACEAE	
Pavonia paniculata	H
Sida sp.	H
MELASTOMATACEAE	
Aciotis sp.	H
OCHNACEAE	
Sauvagesia sp.	H
ONAGRACEAE	
Ludwigia affinis	H
Ludwigia latifolia	H
Ludwigia leptocarpa	H
Ludwigia octovalvis	H
Ludwigia sp.	H
OXALIDACEAE	
Oxalis lespedezioides cf.	H
PASSIFLORACEAE	
Passiflora quinquefolia	V
PHYTOLACCACEAE	
Microtea debilis	H
Phytolacca rivinoides	H
PIPERACEAE	
Pothomorphe peltata	H
POLYGALACEAE	
Polygala acuminata	S
PORTULACACEAE	
Talinum sp.	H

T	tree (dbh 10 cm, height 5 m)
S	shrub
L	liana
V	herbaceous vine
H	herb
E	epiphyte

RUBIACEAE	
Manettia sp.	V
RUTACEAE	
Dictyoloma peruviana	S
SCROPHULARIACEAE	
Lindernia sp.	H
Lindernia crustacea	H
Mecardonia procumbens	H
Scoparia dulcis	H
? sp.	H
SOLANACEAE	
Physalis pubescens	H
Solanum caricaefolium	S
Solanum poeppigianum cf.	S
VERBENACEAE	
Aegiphila integrifolia	S
Lantana trifolia	S
Stachytarpheta sp.	H

Plant List: Apolo Mid-Elevation Wet Forest

Robin B. Foster, Alwyn H. Gentry, Stephen Beck, 1990

Calabatea Forest
Río Yuyu Drainage

ACANTHACEAE	
Aphelandra sp.	S
Hansteinia crenulata	S
Justicia sp.	S
Justicia sp.	S
Mendoncia sp.	L
Ruellia sp.	S
Ruellia sp.	S
Ruellia sp.	S
? sp.	S
ANACARDIACEAE	
Tapirira sp.	T
Tapirira guianensis	T
ANNONACEAE	
Guatteria sp.	T
Guatteria sp.	T
Guatteria sp.	T
Oxandra sp.	T
Xylopia sp.	T
APOCYNACEAE	
Aspidosperma sp.	T
Prestonia sp.	L
ARACEAE	
Anthurium sp.	E
Anthurium sp.	E
Anthurium sp.	E
Anthurium sp.	E
Philodendron ernestii	E
Philodendron sp.	E
Philodendron sp.	E
ARALIACEAE	
Dendropanax sp.	S
Didymopanax morototoni	T
Oreopanax sp.	T
Oreopanax sp.	T
Schefflera sp.	T
Schefflera sp.	T
Schefflera sp.	T
BEGONIACEAE	
Begonia parviflora	S
BIGNONIACEAE	
Anemopaegma sp	L
Arrabidaea patellifera	L
Arrabidaea pearcei	L
Callichlamys latifolia	L
? sp.	L
BORAGINACEAE	
Cordia sp.	T
Cordia sp.	T
BROMELIACEAE	
Tillandsia sp.	E
? sp.	E
BRUNNELIACEAE	
Brunellia sp.	T
BURSERACEAE	
Protium sp.	T
CACTACEAE	
Epiphyllum phyllanthus cf.	E
CHLORANTHACEAE	
Hedyosmum sp.	T
CHRYSOBALANACEAE	
Licania sp.	T
CLETHRACEAE	
Clethra sp.	T
Clethra sp.	T
COMMELINACEAE	
? sp.	H
COMPOSITAE	
Mikania sp.	L
Munnozia sp.	S
Vernonia sp.	S
? sp.	L
CONNARACEAE	
Connarus sp.	L

T	tree (dbh 10 cm, height 5 m)
S	shrub
L	liana
V	herbaceous vine
H	herb
E	epiphyte

Calabatea Forest Río Yuyu Drainage
continued

CONVOLVULACEAE	
Maripa sp.	L
CUNONIACEAE	
Weinmannia sp.	T
CYCLANTHACEAE	
Asplundia sp.	E
DILLENIACEAE	
Doliocarpus sp.	L
Doliocarpus sp.	L
Saurauia sp.	T
? sp.	L
ELAEOCARPACEAE	
Sloanea sp.	T
ERICACEAE	
Cavendishia sp.	E
? sp.	L
ERYTHROXYLACEAE	
Erythroxylum sp.	T
EUPHORBIACEAE	
Alchornea sp.	T
Aparisthmium cordatum	T
Croton sp.	T
Croton sp.	T
Hevea brasiliensis	T
Hyeronima sp.	T
Mabea sp.	T
Maprounea sp.	T
Sapium sp.	T
Tetrorchidium sp.	T
Alchornea trinervis?	T
FLACOURTIACEAE	
Casearia sp.	T
? sp.	T
GENTIANACEAE	
Macrocarpaea sp.	S
Tachia sp.	S
Voyria sp.	H
GESNERIACEAE	
Besleria sp.	S
Besleria sp.	S
GRAMINEAE	
Andropogon bicornis	H
Andropogon sp.	H
Aristida sp.	H
Aristida sp.	H
Axonopus sp.	H
Axonopus sp.	H
Axonopus sp.	H
Brachiaria sp.	H
Chusquea sp.	S
Chusquea sp.	S
Chusquea sp.	S
Loudetia flammida	H
Olyra sp.	H
Panicum sp.	H
Panicum sp.	H
Pariana sp.	H
Schizachyrium condensatum	H
S. microstachyum	H
S. sanguineumcf.	H
Trachypogon plumosus	H
? sp.	H
GUTTIFERAE	
Clusia sp.	T
Clusia sp.	T
Clusia sp.	T
Garcinia(Rheedia) sp.	T
Symphonia globulifera	T
Tovomita weddeliana	T
Vismia sp.	T
ICACINACEAE	
Calatola sp.	T
LABIATAE	
Hyptis hirsuta cf.	S
Hyptis odorata cf.	S
LAURACEAE	

Aniba sp.	T
Endlicheria sp.	T
Licaria? sp.	T
? sp.	T
? sp.	T
? sp.	T
? sp.	T
? sp.	T
? sp.	T
LEGUMINOSAE-CAES	
Swartzia sp.	T
Tachigali sp.	T
LEGUMINOSAE-MIM	
Acacia sp.	L
Inga sp.	T
Inga sp.	T
Inga sp.	T
Inga sp.	T
Parkia sp.	T
Pithecellobium sp.	T
LEGUMINOSAE-PAP	
Lonchocarpus sp.	T
Machaerium sp.	L
Ormosia sp.	T
Platymiscium sp.	T
LINACEAE	
Roucheria sp.	T
LORANTHACEAE	
Gaiadendron sp.	S
Phoradendron sp.	E
Phoradendron sp.	E
Struthanthus sp.	E
MALPIGHIACEAE	
? sp.	L
? sp.	L
? sp.	L
MARANTACEAE	
Calathea sp.	H
Calathea sp.	H
Maranta? sp.	H
Monotagma parvulum cf.	H
MARCGRAVIACEAE	
Marcgravia sp.	L
Marcgravia sp.	L
? sp.	L
? sp.	L
MELASTOMATACEAE	
Aciotis sp.	H
Blakea mexiae cf.	E
Clidemia sp.	S
Clidemia sp.	S
Clidemia sp.	S
Henriettea sp.	T
Henriettella sp.	T
Leandra sp.	S
Miconia sp.	S
Miconia sp.	T
Miconia sp.	T
Miconia sp.	S
Miconia sp.	S
Miconia sp.	T
Miconia sp.	T
Miconia sp.	S
Miconia sp.	S
Tibouchina sp.	H
Tibouchina sp.	T
Tibouchina sp.	S
Topobea sp.	T
Topobea sp.	E
? sp.	S
? sp.	S
? sp.	H
MELIACEAE	
Cabralea cangerana	T
MENISPERMACEAE	
Cissampelos sp.	V
Curarea sp.	L
Orthomene schomburgkii cf.	L
MONIMIACEAE	

T	tree (dbh 10 cm, height 5 m)
S	shrub
T	liana
V	herbaceous vine
H	herb
E	epiphyte

Calabatea Forest Río Yuyu Drainage

continued

Mollinedia sp.	T
Siparuna sp.	S
MORACEAE	
Cecropia sp.	T
Coussapoa sp.	T
Coussapoa sp.	E
Ficus sp.	T
Ficus sp.	T
Ficus sp.	T
Helicostylis sp.	T
Perebea sp.	T
Pourouma minor	T
Pourouma sp.	T
Pseudolmedia sp.	T
Pseudolmedia laevigata cf.	T
Pseudolmedia laevis	T
MUSACEAE	
Heliconia sp.	H
MUSCI	
Sphagnum sp.	H
MYRISTICACEAE	
Virola calophylla cf.	T
MYRSINACEAE	
? sp.	T
Cybianthus sp.	S
Cybianthus sp.	T
Cybianthus sp.	S
Myrsine sp.	T
MYRTACEAE	
Eugenia sp.	T
? sp.	S
? sp.	S
? sp.	T
? sp.	T
NYCTAGINACEAE	
Neea sp.	T
OCHNACEAE	
Ouratea sp.	S
OLACACEAE	
Heisteria sp.	L
ORCHIDACEAE	
? sp.	E
? sp.	E
? sp.	E
? sp.	E
? sp.	E
? sp.	E
PALMAE	
Aiphanes sp.	S
Chamaedorea sp.	S
Dictyocaryum lamarckianum	T
Euterpe sp.	T
Geonoma sp.	S
Iriartea deltoidea	T
Socratea exorrhiza	T
PAPAVERACEAE	
Bocconia sp.	S
PASSIFLORACEAE	
Passiflora sp.	L
PIPERACEAE	
Peperomia sp.	H
Peperomia sp.	E
Peperomia sp.	E
Piper obliqum cf.	S
Piper sp.	S
Piper sp.	S
Piper sp.	L
Piper sp.	S
Piper sp.	S
PODOCARPACEAE	
Podocarpus sp.	T
POLYGALACEAE	
Monnina sp.	S
Polygala gigantea	S
Polygala sp.	H
POLYGONACEAE	

Coccoloba sp.	T
PROTEACEAE	
Roupala montana	T
PTERIDOPHYTA	
? sp.	L
? sp.	H
? sp.	H
? sp.	H
? sp.	E
? sp.	E
? sp.	E
? sp.	E
'treefern' sp.	S
'treefern' sp.	S
'treefern' sp.	S
Asplenium sp.	E
Blechnum sp.	H
Elaphoglossum sp.	E
Elaphoglossum sp.	E
Gleichenia sp.	L
Lindsaea sp.	H
Lycopodium sp.	H
Oleandra sp.	E
Polybotrya sp.	E
Pteridium aquilinum	H
Selaginella sp.	H
Selaginella anceps	H
Selaginella sp.	H
Trichomanes sp.	H
Trichomanes sp.	H
Trichomanes sp.	H
Trichomanes sp.	E
Trichipteris sp.	S
QUIINACEAE	
Quiina sp.	T
ROSACEAE	
Prunus sp.	T
RUBIACEAE	
Bathysa sp.	T

Cephaelis sp.	S
Cephaelis tomentosa	S
Cephaelis ulei?	S
Chomelia sp.	L
Coccosypselum sp.	H
Coussarea sp.	T
Coussarea sp.	T
Faramea sp.	S
Faramea sp.	T
Hillia sp.	E
Ladenbergia? sp.	T
Ladenbergia? sp.	T
Ladenbergia? sp.	T
Palicourea sp.	S
Psychotria officinalis cf.	S
Psychotria orchidearum	S
Psychotria sp.	S
Psychotria sp.	S
Psychotria sp.	S
Psychotria sp.	S
Psychotria sp.	S
Rudgea sp.	T
? sp.	T
? sp.	T
? sp.	T
? sp.	T
? sp.	T
RUTACEAE	
Zanthoxylum sp.	T
SABIACEAE	
Meliosma? sp.	T
SAPINDACEAE	
Paullinia sp.	L
Paullinia sp.	L
SAPOTACEAE	
Pouteria sp.	T
Pouteria sp.	T
Pouteria sp.	T
? sp.	T
SCROPHULARIACEAE	

T	tree (dbh 10 cm, height 5 m)
S	shrub
T	liana
V	herbaceous vine
H	herb
E	epiphyte

Calabatea Forest
Río Yuyu Drainage
continued

? sp.	H
SIMAROUBACEAE	
Picramnia sp.	T
Simarouba amara	T
SOLANACEAE	
Brunfelsia sp.	S
Cestrum sp.	S
Juanulloa sp.	E
Markea sp.	E
Solanum anceps	S
Solanum argenteum	S
Solanum sp.	T
SYMPLOCACEAE	
Symplocos sp.	T
THEACEAE	
Laplacea sp.	T
UMBELLIFERAE	
Hydrocotyle sp.	H
URTICACEAE	
Pilea sp.	H
VIOLACEAE	
Leonia glycicarpa	T
VOCHYSIACEAE	
Erisma? sp.	T
ZINGIBERACEAE	
Renealmia thyrsoidea	H

Apolo Mattorral

ANACARDIACEAE	
? sp.	T
ARALIACEAE	
Didymopanax morototoni	T?
ASCLEPIADACEAE	
? sp.	H
BURMANNIACEAE	
Apteria aphylla	H
CAMPANULACEAE	
Centropogon sp.	H
COMPOSITAE	
Chromolaena sp.	S
Eupatorium sp.	S
Eupatorium sp.	S
Lycoseris sp.	V
Mikania sp.	L
Vernonia sp.	S
? sp.	S
? sp.	S
? sp.	S
? sp.	H
CYPERACEAE	
Bulbostylis sp.	H
Eleocharis sp.	H
Fimbristylis sp.	H
Rhynchospora rugosa	H
Scleria sp.	H
Scleria sp.	H
EUPHORBIACEAE	
Alchornea triplinervia	T
GENTIANACEAE	
Curtia sp.	H
GRAMINEAE	
Andropogon sp.	H
Aristida capillacea cf.	H
Aristida sp.	H
Axonopus sp.	H
Brachiaria sp.	H
Chusquea sp.	H
Eragrostis sp.	H
Panicum sp.	H
Panicum stenoides cf.	H
Paspalum sp.	H
Paspalum sp.	H
Schizachyrium sp.	H
Schizachyrium sp.	H
? sp.	H

APPENDIX 8

Species	
LABIATAE	
Hyptis sp.	S
Hyptis sp.	H
LEGUMINOSAE-PAP	
Desmodium barbatum cf.	H
Stylosanthes sp.	H
MALPIGHIACEAE	
Byrsonima sp.	T
? sp.	S
MELASTOMATACEAE	
Desmoscelis villosa	S
Miconia sp.	S
Miconia sp.	S
Miconia sp.	S
Miconia sp.	S
Miconia albicans	S
Miconia rufescens	S
Tibouchina sp.	S
MYRTACEAE	
Psidium guajava?	S
? sp. a	S
? sp. b	S
OCHNACEAE	
Sauvagesia sp.	H
PTERIDOPHYTA	
Gleichenia sp.	H
RUBIACEAE	
Sabicea pedunculata	L
RUTACEAE	
Dictyoloma peruviana	S
SCROPHULARIACEAE	
Buchnera sp.	H
SOLANACEAE	
Solanum sp.	S
THEACEAE	
Ternstroemia sp.	T

T	tree (dbh 10 cm, height 5 m)
S	shrub
T	liana
V	herbaceous vine
H	herb
E	epiphyte

APPENDIX 9

Plant List: Apolo - Mid-Elevation Dry Forest

Robin B. Foster, Alwyn H. Gentry, 1990

Deciduous Forest Río Machariapo

ACANTHACEAE	
Aphelandra sp.	S
Justicia sp.	H
Justicia sp.	H
Justicia sp.	S
Ruellia sp.	S
Sanchezia sp.	H
AMARANTHACEAE	
Alternanthera sp.	H
Amaranthus? sp.	H
Celosia sp.	H
Chamissoa sp.	V
Iresine sp.	L
? sp.	H
AMARYLLIDACEAE	
? sp.	H
ANACARDIACEAE	
Astronium sp.	T
Schinopsis sp.	T
APOCYNACEAE	
Aspidosperma sp.	T
Forsteronia sp.	L
Forsteronia spicata	L
Prestonia sp.	L
Prestonia sp.	L
Tabernaemontana sp.	S
Tabernaemontana sp.	T
ARACEAE	
Anthurium clavigerum	E
Anthurium sp.	H
Anthurium sp.	H
Anthurium sp.	E
Philodendron sp.	E
ASCLEPIADACEAE	
Matelea sp.	V
? sp.	V

BEGONIACEAE	
Begonia sp.	H
BIGNONIACEAE	
Amphilophium paniculatum	L
Arrabidaea conjugata	L
Arrabidaea corallina	L
Arrabidaea poeppigii	L
Arrabidaea selloi	L
Callichlamys latifolia	L
Clytostoma uleanum	L
Macfadyena unguiscati	L
Mansoa difficilis	L
Melloa quadrivalvis	L
Paragonia pyramidata	L
Pithecoctenium crucigerum	L
Tabebuia impetiginosa	T
Tabebuia ochracea	T
Tabebuia serratifolia	T
BOMBACACEAE	
Ceiba sp.	T
Ochroma pyramidale	T
? sp.	T
BORAGINACEAE	
Cordia alliodora aff.	T
BROMELIACEAE	
Aechmea sp.	E
Bromelia? sp.	H
Tillandsia sp.	E
Tillandsia sp.	H
? sp.	H
? sp.	H
CACTACEAE	
Acanthocereus? sp.	T
Cereus sp.	T
Epiphyllum phyllanthus cf.	E
Hylocereus sp.	E
Opuntia sp.	S
Pereskia sp.	T

Species	Habit
? sp.	S
CAPPARIDACEAE	
Capparis sp.	T
CELASTRACEAE	
Maytenus sp.	T
COMBRETACEAE	
Combretum sp.	L
Combretum sp.	L
? sp.	S
COMMELINACEAE	
Campelia zanonia	H
? sp.	V
COMPOSITAE	
Verbesina sp.	S
CONNARACEAE	
Connarus sp.	L
CONVOLVULACEAE	
Ipomoea sp.	V
? sp.	L
CUCURBITACEAE	
Psiguria sp.	V
DIOSCOREACEAE	
Dioscorea sp.	V
EUPHORBIACEAE	
Acalypha sp.	S
Acalypha sp.	S
Cnidoscolus sp.	S
Croton sp.	T
Euphorbia sp.	S
? sp.	S
FLACOURTIACEAE	
Casearia sylvestris	T
Xylosma sp.	T
GRAMINEAE	
Chusquea sp.	S
Olyra sp.	H
GUTTIFERAE	
Clusia sp.	E

Species	Habit
HIPPOCRATEACEAE	
Hippocratea sp.	L
LAURACEAE	
Ocotea sp.	T
? sp.	T
LEGUMINOSAE-MIM	
Acacia sp.	T
Acacia sp.	T
Acacia sp.	L
Anadenanthera colubrina	T
Inga sp.	T
Inga edulis cf.	T
Piptadenia comunis	L
Piptadenia flava cf.	L
LEGUMINOSAE-PAP	
Amburana cearensis	T
Caesalpinia? sp.	T
Dalbergia sp.	L
Lonchocarpus sp.	T
Machaerium sp.	T
Machaerium sp.	L
Machaerium sp.	L
Myroxylon balsamum	T
Platymiscium sp.	T
Pterocarpus sp.	T
? sp.	T
LORANTHACEAE	
Phoradendron sp.	E
MELIACEAE	
Cedrela sp.	T
Trichilia elegans	T
Trichilia sp.	S
Trichilia sp.	T
MORACEAE	
Cecropia sp.	T
Cecropia polystachya	T
Clarisia biflora	T
Ficus citrifolia	T
Ficus juruensis	E
Maclura (Chlorophora) tinctoria	T

T	tree (dbh 10 cm, height 5 m)
S	shrub
L	liana
V	herbaceous vine
H	herb
E	epiphyte

Deciduous Forest Río Machariapo
continued

NYCTAGINACEAE	
Neea sp.	S
Pisonia sp.	T
? sp.	S
? sp.	T
OPILIACEAE	
Agonandra sp.	T
ORCHIDACEAE	
? sp.	H
? sp.	E
? sp.	E
? sp.	E
? sp.	E
? sp.	E
? sp.	E
? sp.	H
? sp.	E
? sp.	E
PALMAE	
Syagrus sp.	T
PHYTOLACCACEAE	
Achatocarpus sp.	T
Gallesia integrifolia	T
Petiveria alliacea	S
Seguieria macrophylla	L
PIPERACEAE	
Peperomia sp.	H
Peperomia sp.	H
Piper sp.	S
Piper medium	S
POLYGONACEAE	
Triplaris sp.	T
PORTULACACEAE	
Portulaca? sp.	H
PTERIDOPHYTA	
Platycerium andinum	E
Polypodium sp.	E
RHAMNACEAE	
Gouania sp.	L
RUBIACEAE	
Pittoniotis sp.	T
Randia sp.	S
SAPINDACEAE	
Allophylus sp.	T
Sapindus saponaria	T
Serjania sp.	L
Serjania sp.	L
Thinouia sp.	L
Urvillea? sp.	L
SAPOTACEAE	
Pouteria sp.	T
Pradosia sp.	T
SOLANACEAE	
Solanum sp.	S
STERCULIACEAE	
Guazuma ulmifolia	T
Helicteres sp.	S
THEOPHRASTACEAE	
Clavija sp.	S
TILIACEAE	
Luehea grandiflora cf.	T
TRIGONIACEAE	
Trigonia sp.	L
ULMACEAE	
Ampelocera sp.	T
Celtis iguanea	L
Celtis sp.	L
Phyllostylon sp.	T
URTICACEAE	
Urera baccifera	S
Urera caracasana	S

Montane Savanna Chaquimayo

ANACARDIACEAE	
Schinopsis sp.	T
? sp.	T
BIGNONIACEAE	
Jacaranda cuspidifolia	T
Tabebuia aurea	T
Tabebuia roseo-alba	T
BOMBACACEAE	
Pseudobombax sp.	T
CAPRIFOLIACEAE	
Viburnum sp.	S
COMBRETACEAE	
Terminalia sp.	T
COMPOSITAE	
? sp.	H
EBENACEAE	
Diospyros sp.	T
GRAMINEAE	
? sp.	H
? sp.	H
IRIDACEAE	
Sisyrinchium sp.	H
LEGUMINOSAE-CAES	
Senna sp.	T
LEGUMINOSAE-PAP	
Machaerium sp.	L
LORANTHACEAE	
Struthanthus sp.	E
MALPIGHIACEAE	
Mascagnia sp.	L
MELASTOMATACEAE	
? sp.	S
ORCHIDACEAE	
? sp.	E
PROTEACEAE	
Roupala sp.	T
RUBIACEAE	
Condaminea corymbosa	S
SAPINDACEAE	
Dilodendron bipinnatum	T
Serjania sp.	L
STYRACACEAE	
Styrax sp.	S
TILIACEAE	
Luehea sp.	S

T	tree (dbh 10 cm, height 5 m)
S	shrub
L	liana
V	herbaceous vine
H	herb
E	epiphyte

APPENDIX 10

Plant List: Pampa - Ixiamas

Robin B. Foster, Alwyn H. Gentry, StephanBeck, 1990

ACANTHACEAE	
Mendoncia sp.	L
Staurogyne diantheroides cf.	H
ALISMATACEAE	
Sagittaria guyanensis	H
ANNONACEAE	
Xylopia frutescens	T
APOCYNACEAE	
Himatanthus sp.	T
Mandevilla sp.	V
BIGNONIACEAE	
Arrabidaea sp.	L
Ceratophytum tetragonolobum	L
Paragonia pyramidata	L
Pithecoctenium crucigerum	L
Tabebuia ochracea	T
Tabebuia serratifolia	T
BOMBACACEAE	
Chorisia sp.	T
Pseudobombax sp.	T
Pseudobombax sp.	T
Cordia sp.	S
BURMANNIACEAE	
Burmannia capitata	H
Burmannia sp.	H
Burmannia sp.	H
Burmannia sp.	H
BURSERACEAE	
Protium sp.	T
CACTACEAE	
Pereskia sp.	S
CAMPANULACEAE	
Lobelia sp.	H
CARYOPHYLLACEAE	
? sp.	H
CHRYSOBALANACEAE	
Hirtella sp.	S
COCHLOSPERMACEAE	
Cochlospermum vitifolium	T
COMPOSITAE	
? sp.	H
Ayapana amygdalina	H
Baccharis chilca	S
Calea sp.	S
Clibadium sp.	S
Conyza sp.	H
Eupatorium sp.	H
Eupatorium sp.	S
Mikania officinalis	H
Vernonia baccharoides cf.	S
Vernonia macrophylla	H
Vernonia sp.	S
CONVOLVULACEAE	
Cuscuta sp.	H
CYCADACEAE	
Zamia boliviana	S
CYPERACEAE	
? sp.	H
Bulbostylis junciiformis cf.	H
Cyperus haspan	H
Cyperus sp.	H
Eleocharis sp.	H
Eleocharis sp.	H
Fuirena robusta cf.	H
Rhynchospora globosa cf.	H
Scleria hirtella cf.	H
Scleria natans cf.	H
Scleria sp.	H
DILLENIACEAE	
Curatella americana	T
DROSERACEAE	
Drosera sp.	H
ERIOCAULACEAE	
Syngonanthus aulescens	H
Syngonanthus gracilis	H
Syngonanthus sp.	H
EUPHORBIACEAE	

APPENDIX 10

? sp.	H
Caperonia palustrus	H
Caperonia sp.	H
GENTIANACEAE	
Curtia tenella	H
Schultesia sp.	H
GRAMINEAE	
? sp.	H
? sp.	H
? sp.	H
? sp.	H
Andropogon sp.	H
Aristida capillacea cf.	H
Aristida sp.	H
Aristida sp.	H
Axonopus sp.	H
Hemarthria altissima	H
Hyperrhenia bracteata	H
Loudetia sp.	H
Panicum sp.	H
Panicum sp.	H
Panicum stenoides cf.	H
Paspalum sp.	H
Paspalum sp.	H
Sacciolepis sp.	H
Sacciolepis sp.	H
Sacciolepis sp.	H
Schizachyrium sanguineum	H
Schizachyrium sp.	H
Schizachyrium sp.	H
Schizachyrium sp.	H
Sorghastrum stipoides cf	H
Trachypogon plumosus	H
HYDROPHYLLACEAE	
Hydrolea sp.	H
LABIATAE	
Hyptis carpinifolia	S
Hyptis sp.	H
Hyptis sp.	S
Hyptis sp.	S
Hyptis sp.	H
LEGUMINOSAE-CAES	
Bauhinia sp.	S
Chamaecrista sp.	H
LEGUMINOSAE-PAP	
? sp.	L
Calapogonium sp.	L
Crotalaria sagittata	H
Crotalaria sp.	H
Crotalaria sp.	H
Desmodium triflorum	H
Eriosema sp.	S
Eriosema sp.	S
Indigofera lespedezioides	S
Machaerium sp.	T
Stylosanthes sp.	H
LIMNOCHARITACEAE	
Hydrocleys sp.	H
Limnocharis sp.	H
LYTHRACEAE	
Cuphea sp.	H
Cuphea sp.	H
MALPIGHIACEAE	
Stigmaphyllon sp.	L
MALVACEAE	
Abelmoschus? sp.	S
Hibiscus? sp.	S
Peltaea sp.	S
MELASTOMATACEAE	
Aciotis sp.	H
Clidemia sp.	H
Desmoscelis sp.	H
Rhynchanthera sp.	H
Siphanthera? sp.	H
MELIACEAE	
Guarea sp.	T
MONIMIACEAE	
Siparuna sp.	T
Siparuna sp.	T

T	tree (dbh 10 cm, height 5 m)
S	shrub
L	liana
V	herbaceous vine
H	herb
E	epiphyte

MORACEAE	
Cecropia sp.	T
MUSACEAE	
Heliconia sp.	H
MYRISTICACEAE	
Virola sebifera	T
OCHNACEAE	
Sauvagesia nana	H
Sauvagesia sp.	H
ONAGRACEAE	
Ludwigia sp.	H
OXALIDACEAE	
Oxalis sp.	S
PALMAE	
Allagoptera leucocalyx	S
Mauritia flexuosa	T
PIPERACEAE	
Piper sp.	S
Piper sp.	S
POLYGALACEAE	
Polygala asperuloides	H
Polygala sp.	H
Polygala sp.	H
Polygala timontoides cf.	H
PRIMULACEAE	
Anagallis pumila	H
PTERIDOPHYTA	
Lycopodiella sp.	H
Selaginella sp.	H
RUBIACEAE	
Coussarea sp.	H
Diodia sp.	H
Diodia sp.	H
Genipa americana	T
Psychotria sp.	S
Sabicea sp.	L
SAPINDACEAE	
Cupania sp.	T
SCROPHULARIACEAE	
? sp.	H
Bacopa sp.	H
Buchnera juncea	H
Melasma? sp.	H
SMILACACEAE	
Smilax sp.	L
SOLANACEAE	
Brunfelsia sp.	S
Cyphomandra sp.	S
STERCULIACEAE	
Byttneria sp.	S
Helicteres sp.	S
TILIACEAE	
Corchorus sp.	H
Luehea sp.	T
Triumfetta sp.	S
UMBELLIFERAE	
Eryngium elegans	H
VERBENACEAE	
Lippia vernonioides	S
VITACEAE	
Cissus sp.	L
XYRIDACEAE	
Xyris sp.	H
Xyris sp.	H
Xyris sp.	H
Xyris sp.	H
Xyris sp.	H
Xyris sp.	H

Bibliography

Davis, T.H. 1986. Distribution and natural history of some birds from the departments of San Martin and Amazonas, northern Peru. Condor 88:50-56.

Graves, G.R. and R.L. Zusi. 1990. Avian body weights from the lower Rio Xingu, Brazil. Bull. Brit. Orn. Cl. 110:20-25.

Haase, R. and S. Beck. 1989. Structure and composition of savanna vegetation in northern Bolivia: a preliminary report. Brittonia 41(1): 80-100.

O'Neill, J.P., C.A. Munn, and I. Franke J. 1991. *Nannopsitacca dachileae*, a new species of parrotlet from eastern Peru. Auk 108:225-229. [cited as O'Neill et. al. in press in text]

Parker, T.A., III. 1982. Observations of some unusual rainforest and marsh birds in southeastern Peru. Wilson Bull. 94:477-493.

Parker, T.A., III, and J.V. Remsen, Jr. 1987. Fifty-two Amazonian bird species new to Bolivia. Bull. Brit. Orn. Cl. 107:94-107.

Parker, T.A., III, and S. A. Parker. 1982. Behavioral and distributional notes on some unusual birds of a lower montane cloud forest in Peru. Bull. Brit. Orn. Cl. 102:63-70.

Pearson, D.L., D. Tallman, and E. Tallman. 1977. The birds of Limoncocha, Napo Province, Ecuador. Instituto Linguistico de Verano, Quito (privately published).

Remsen, J.V., Jr. 1986. Ecological profile of a lower montane cloudforest in northern Bolivia. MS.

Robbins, M.B., R.S. Ridgely, T.S. Schulenberg, and F.B. Gill. 1987. The avifauna of the Cordillera de Cutucu, Ecuador, with comparisons to other Andean localities. Proc. Acad. Nat. Sci. Philadelphia 139:243-259.

Robbins, M.B., S. Cardiff, A. Capparella, and R.S. Ridgely. 1992. The avifauna of the Rio Manati and Quebrada Vainilla areas in northeastern Peru. Proc. Acad. Nat. Sci. Phiad. (in press).

Schulenberg, T.S., S.E. Allen, D.F. Stotz, and D.A. Wiedenfeld. 1984. Distributional records from the Cordillera Yanachaga, central Peru. Gerfaut 74:57-70.

Stotz, D.F. and R.O. Bierrgaard, Jr. 1989. The birds of the Fazendas Porto Alegre, Esteio and Dimona north of Manaus, Amazonas, Brazil. Rev. Brasil. Biol. 49:861-872.

Terborgh, J. and J.S. Weske. 1979. The role of competition in the distribution of Andean birds. Ecol. 56:562-576.

Terborgh, J.W., J.W. Fitzpatrick, and L. Emmons. 1984. Annotated checklist of bird and mammal species of Cocha Cashu Biological Station, Manu National Park, Peru. Fieldiana (Zoology, New Series) 21:1-29.

Terborgh, J., S.K. Robinson, T.A. Parker, III, C.A. Munn, and N. Pierpont. 1990. Structure and organization of an Amazonian forest bird community. Ecol. Monogr. 60:213-228.

Willis, E.O. 1977. Lista preliminar das aves da parte noroeste e areas vizinhas de Reserva Ducke, Amazonas, Brazil. Rev. Brasil. Biol. 37:585-601.